What people are saying about

# Spheres of Perception

Moving beyond and between disciplines and the effects of technology on our lives, this book provides a rich and sophisticated transdisciplinary exploration of humanity's 'being in this world.' The reflections on our logical, physical, and metaphysical evolution challenge our illusions about humanity's competence to overcome disparities between the way we live and the way we develop. This book must be read by everybody looking for a sensible and holistic evaluation of the drastic challenges we face and the transformations we require to adapt to the present.

**Dr Hester du Plessis,** DLitt et Phil, Chief Research Specialist, Human Sciences Research Council (HSRC), South Africa

# Spheres of Perception

## Morality in a Post Technocratic Society

Theodore Holtzhausen

# Spheres of Perception

## Morality in a Post Technocratic Society

Theodore Holtzhausen

Winchester, UK
Washington, USA

JOHN HUNT PUBLISHING

First published by Changemakers Books, 2020
Changemakers Books is an imprint of John Hunt Publishing Ltd., No. 3 East Street,
Alresford, Hampshire SO24 9EE, UK
office@jhpbooks.com
www.johnhuntpublishing.com
www.changemakers-books.com

For distributor details and how to order please visit the 'Ordering' section on our website.



ISBN: 978 1 78535 795 4
978 1 78535 892 0 (ebook)
Library of Congress Control Number: 2017957125



A CIP catalogue record for this book is available from the British Library.

Design: Stuart Davies

**Disclaimer**: The novel concept and conclusions presented in this book are those of the author. It is not his intention to either unnecessarily duplicate or misrepresent any preexisting concepts on these matters that may exist or have acted in support of this text.

UK: Printed and bound by CPI Group (UK) Ltd, Croydon, CR0 4YY
US: Printed and bound by Thomson-Shore, 7300 West Joy Road, Dexter, MI 48130

# Contents

**Also by Theodore Holtzhausen**

*Sensible Gene Selfish Being*

ISBN 978-981-08-7039-3

This book is dedicated to all those who without exploitation quietly, unnoticed, and selflessly help to evolve our knowledge and morality, to improve the world for present and future generations.

A special thank-you goes to my publisher, Tim Ward, without whom I could not have completed this project, in the end a larger task than we both thought.

Recent findings in the biomedical and physical sciences have dramatically changed the way we see the world and our place in it. This significant paradigm shift in our understanding will dramatically affect every part of our future existence. Strangely, this knowledge has not sufficiently infiltrated our socioeconomical structures yet. Eminent and urgent, then, has become the need to *single out* and update our outmoded economic and healthcare systems. Both current arrangements are dismally failing to meet the demands set by our rapidly evolving epistemology. A progressive new model is presented here.

> I shall never rest until I know that all my ideas are derived, not from hearsay or tradition, but from my real living contact with the things themselves.
>
> Goethe, *Italian Journey* (1816–17)

## Abbreviations and formulae introduced in this text:

PSR Physical sphere of reasoning
LSR Logical sphere of reasoning
MS Metaphysical sphere
M Metaphysical
PMD Physical sphere/Metaphysical Dilemma
DNA Deoxyribonucleic acid
RNA Ribonucleic acid
TT Tentative theory
EE Error elimination
P1 Problem 1
P2 Problem 2
QOL Quality of Life

$\infty\Delta a \approx \infty\Delta b$ where *a* is the observer and *b* is the observed
Ev Evolution
Ev(mo) Moral evolution
C Cognition

Traditional natural selection = $\Delta a(C) \leftrightharpoons \Delta b$

*or*

Perceptive Ev(mo) = $\sum \infty\Delta C\{\infty\Delta a(\text{Metaphysical} \leftrightharpoons \text{LSR} \leftrightharpoons \text{PSR}) \approx \infty\Delta b(\text{Metaphysical} \leftrightharpoons \text{LSR} \leftrightharpoons \text{PSR})\}$

**Illustrations:**

Figures 1, 2 and 4 © author
Figure 3. Citric Acid Cycle (Ophardt, 2003)

# 1

# Preface

*The most important human endeavor is the striving for morality in our actions. Our inner balance and our very existence depend on it. Only morality in our actions can give beauty and dignity to life.*
Albert Einstein

*In considering an ethos in healthcare securing its foundations in pragmatic and pure knowledge free of pseudoscience and fraud, we should foremost ask ourselves how to safeguard it against delusional belief systems and impetuous profiteering.*
Author, from abstract delivered in Prague, 2014

## 1.1 Introduction

We live in an astonishing era with an unprecedented dependency on modern technology. A constant flow of new ideas and opinions bombards us daily while our eyes are glued to our various electronic devices. A new electronic battleground has emerged to influence what people see, believe and think. With much of our information production motivated by profits or self-interest, our collective knowledge is becoming undeniably biased—at times even false, as we lack the filters in our electronic systems to weed out the lies and prevent social media from spreading them around the world. At the same time, we are suffering information overload. We are bombarded by advertisements and social media marketing that seeks to draw our attention and sway our opinion. How are we to determine what new information is accurate and important in the midst of this barrage?

Science, meanwhile, has been making discoveries that would dramatically change the way we see and understand ourselves—if that new information could get through the blaring noise. Instead, there is an increasing discrepancy between what science knows and what our current economic system tells us is true. This is most pointedly true when it comes to the study of evolution. Science is discovering our place in the universe as much more interconnected and interactive than previously recognized, while the current economic system in turn, by design, fosters exclusion and segregation. This is a significant shift in our evolutionary paradigm; yet the importance of this metamorphosis has not spread around the world, and remains largely unknown. Perhaps this is because science doesn't create marketing campaigns, and doesn't have billion-dollar advertising budgets.

Meanwhile, our economic system is becoming over-stimulated by the information age. Interconnection has helped companies earn trillions and rise swiftly to global dominance. But the 24-hour wired world has also led to increased volatility, and negative information; even an accidental computer glitch can plunge the market and create panic. Corporations must pay ever more attention to the short-term bottom line. Shareholder profits, at all cost, seem to be what matters most. Our economies are therefore based on what can sell, rather than what can genuinely improve the human condition. Health, the environment, the welfare of society—these are pushed to the edge of our national concerns. Politicians hand out business incentives and tax breaks, then tell us there's not enough revenue to improve health services. A company wants to build a pipeline, and any wilderness in its path that will be spoiled is just the cost of progress. Valuing short-term profits and growth over long-term impact on society and the environment inevitably will lead to collapse.

On a personal scale, even buying a vacuum cleaner is difficult—with many choices, financing options, warnings and warranties, and information online about each product. Large corporations have now obtained significant power to sway research with bias-targeting profits based on their own interests and world-views. It is now commonplace for the pragmatic value to consumers and the potential negative impact on the environment to be heavily manipulated by clever marketing strategies. In fact, the words 'genuine' and 'truth,' 'evidence-based' and 'peer-reviewed,' have never been more equivocal and potentially more malleable by those with legal and financial muscle. Such ambiguity and manipulations are also threatening to divert a pragmatic, open and truthful science into a misdirected pseudoscience, with the potential to turn our entire evolving epistemology into a misguided fallacious and embarrassing fabrication.

For example, in our heavily corporate-infiltrated healthcare systems, this surge of profit-driven knowledge makes it difficult for the clinician to distribute bona fide and wise treatments to their patients (now called 'consumers of healthcare'). For those seeking treatment, it has become more complex to weigh up the dependability of costly medicines and procedures against their quality of life. With pharmaceutical companies and big corporations operating on a different level, well removed from the emotional impact of disease and suffering, they see any equivocality or falsifiability in a science and its knowledge as new potential for exploitation to maximize profits. This power to sway outcomes is then vulnerable to biases, personal and set worldviews. Subsequently, marketing products with gnomic value backed by pseudoscience are constantly slipping through the system while ethical decision-making in healthcare is growing in complexity.

Healthcare is just one example. It seems that this is just the way the world is; the machine has grown too big, too powerful, too fast for us to change. How can scientists, visionaries and those who care about future generations make their voices heard? What can we do to challenge the primacy of our economic system, and place new emphasis on creating a virtuous society?

I believe an important part of the answer, a part we have neglected thus far, is to develop a trustworthy epistemology. By this I mean a new way of thinking about the knowledge we develop, debate, and disseminate to others. If we can learn to think clearly and act wisely, we may discover a universal morality is within our reach. This is the purpose of this book: to help clarify our ability to think by introducing three **spheres of reasoning**. This, I will argue, is the missing ingredient. If we can possess such a trustworthy epistemology, then our innovative technology plus the findings of scientific research can lead us, perhaps, to understanding our universal evolutionary purpose for the very first time.

This is not at all a theoretical exercise, but rather, a pressing need: a need for concrete and trustworthy guidelines for our society. Specifically, we need something to replace the 'survival of the fittest' approach, both in business and in our daily lives. This mantra of Charles Darwin no longer reflects current evolutionary thinking. Yet this 'survivalist' mentality is partially responsible for making the world an unnecessarily unjust and a much harsher place than it needs to be. Indeed, it legitimizes greed, corruption, and manipulative behavior.

However, the science of evolution has itself evolved. In as much as 'survival of the fittest' has been used to justify harsh, competitive behavior on the part of individuals and corporations, an updated understanding of evolution could lead us to update our ethics. What if written into the code of our DNA and RNA is a guide for how we should behave and live with one another?

Science has come to see life as evolving through responsive and pliable RNA and DNA molecules. They interact in interconnected ways, using various chemical elements and molecules as means of communicating information. In other words, DNA and RNA act as if they are perceiving their environment (and each other), and then communicating about it to each other. This enables them to collaborate on a goal, such as building a specific protein. This is evidence of a more percipient and *mobile* DNA/RNA than previously thought. It has not only dramatically changed the way we understand evolutionary biology, but also has implications for human morality: Connection, communication, and collaboration are in the building blocks of our molecular structure. At the very minimum this calls for a re-evaluation of our reductionist interpretation of evolutionary biology. It needs to be updated and recognized as the *perceptive* process it really is.

No doubt it sounds strange to speak of molecules as perceiving, communicating, and collaborating. We don't think of these as attributes of *things*, of bits of matter. Instead, we think of these as attributes of conscious beings. But is that necessarily so, or is that the materialist Newtonian paradigm lodged so deeply in our minds that it is hard to imagine anything else? In fact, we know that paradigm was wrong about atoms. Einstein showed us that matter is only energy. As quantum physics has shown us, the atom—once thought of as a solid little ball circled by whizzing electrons—is actually a cloud of quarks and particles that themselves dissolve. In reality, we can't really get our minds around what an atom is. Similarly, science has a great deal of difficulty grasping the nature of consciousness. Where exactly is consciousness located? How does it come into being? How does it move matter such that we can intend our hand to open and it does? Why should our human experience of consciousness be the only standard? A dolphin, a bee, a tree—all perceive, communicate, and collaborate. (It's recently been discovered that trees send chemical messages to each other through underground networks of fungi. If these creatures are in some sense conscious, why not DNA?) The only thing we can say for sure about consciousness is that it arises within a body made up entirely of molecules. So, if we are conscious, and all we are made of is molecules, then the rather inescapable conclusion is there must be something in the nature of molecules that enables consciousness in us. Fortunately, it's not the purpose of this book to convince the reader that this is so. It is merely a useful thought-exercise, however, to explore how adherence to dogmatic ideas (reductionist materialism) can block one's openness to logical reasoning (that molecules may have some kind of consciousness). So, when the text refers to molecules as "perceiving," having "concerns," or an "ethic" governing their activities, please remember I'm not implying they have human-like consciousness; but I am using these words as shorthand for describing behavior among molecules that a reductionist model can't easily explain.

Just as quantum physics has shown us that materialism is inadequate (material is an illusion of energy), so modern molecular science is revealing that 'survival of the fittest' is also inadequate. Perceptivity, responsiveness, and collaboration are essential behaviors for evolutionary success. If indeed these are the principles that make life work, then a single-minded focus on competition—a 'survivalist' mentality—is a dangerous delusion that hampers progress and may lead to our extinction. It is made all the worse by the fact that those corporate elements of our society who have embraced a survivalist mentality are the dominant voices in the production and dissemination of knowledge.

In a society that is technologically advanced, yet morally mediocre and dominated by profit-seekers, how can we learn to think clearly? We can now assert that our thoughts and ideas, including our ideas about morality, are no more than *a subset of progressive interconnected evolutionary processes*. We can now also create scope for ongoing adaptability for both how we think and how we behave. This could give us the ability to create a higher level of morality than humanity has ever experienced before, an *internal* evolving morality that is literally in our genes, and that we have transgressed to our peril. As we learn more about evolutionary processes at the molecular level, in the principles behind them we may find further guides towards a more tolerant, respectful, interconnected, and moral society. This path also opens up a new *metaethics*—a way to think about what morality is—and thus gives us a process for continued moral development. The evidence for this will be explained in following chapters.

* * *

The immediate challenge we face is to articulate any realistic and universal opinions on morality and ethics free of biases, for instance personal interests, religious or cultural factors, and politically driven motives. The Harvard logician and Kant scholar Clarence Irving Lewis (1883–1964) proposed that what is right and wrong might be evaluable in terms of whether they fit with experience and survive scrutiny. I see in this pragmatic approach the criteria that ethics be backed by justifiable universal rules supported by both evidence and experience. These rules should be detached from personal reward and also be capable of pragmatic adjustment to meet our evolving needs. Where can we turn to for the kind of experience that will serve as a foundation? Human society (and current ethical systems) may not be the best place to look for ethical norms. Our existing world-views are so biased by prejudice, greed, oppressive power structures, and mistaken and conflicting mythological beliefs, that we need a clean slate to begin. An honest natural science gives us a perspective of 'what is'—provided we can decontaminate it of our biases. Just as scientists are trained to craft experiments in such a way as to avoid observer biases, so too our methods must prevent the biases of false beliefs from creeping in.

This quest is more urgent than most of us realize. Saving the planet may sound a bit grandiose, but in an era of genomics, robotics, and climate change, reviving our moral duty backed by a truthful science can no longer be ignored. It is vital to our ongoing evolution as moral and perceptive beings. Indeed, in the final chapters we will argue that morality and perceptivity are intrinsically entwined.

Here's how I perceive our current ethical situation: We have inherited a diverse set of moral codes that are part of religious belief systems. Mostly these are based on some version of God or gods handing to humans a code of conduct. A good metaphor for this is that God, our maker, has also given us an instruction manual (moral code) for our smooth operation in society. This set-up nicely nestled our morality inside our metaphysics. But the whole package is externally imposed—that is, it is derived from a source (God) outside of ourselves. As science has come to question the metaphysical validity of religion, the ethical foundations nestled within have crumbled too. As a result, humanity finds itself struggling to hold on to a sense of morality in a reductionist world with no external standard of right and wrong. In that world, we have been told that 'survival of the fittest' is what is in our genetic blueprint. Our ethical struggle is between a set of beliefs we can no longer believe in and a grim amoral reality. Our materialistic metaphysics excludes God, and *therefore* gives us no moral code.

This predicament has enabled the rise of our 'survival of the fittest' economic system. Ethical complaints seem like quaint throwbacks to our religious past: unrealistic objections to the way the world really is. One response to this has been to create a morality based on human rights: what we can agree on to value in each other. In other words, not a code God gives us, but rather, a code that we give to one another. While there may be much to admire in this humanist approach, it remains an extrinsic morality: a code imposed upon us, rather than derived from who we truly are. So long as 'survival of the fittest' is seen as the code written in our DNA, it will likely remain a more powerful justification for how we treat each other.

I would argue that we can't advance our morality by ignoring the metaphysical, and that evolutionary science can help us succeed where both religious belief systems and humanism have failed. We must express grave concerns with any search that, firstly, presents a normative, such as a traditional belief, fixed law, or set theory on how to behave, and attempts to define and enforce morals from such an intransigent normative. And then subsequently continues, through power struggles, to attempt to formulate an ethic from this with disregard for its origins in nature, where everything is constantly evolving and adjusting. Such an approach dismally fails to address the changes it has to confront and adjust to as an *evolving interconnected* perceptive network. With such an approach, the end-product would also be subjective, equivocal, and not practical or universally applicable. Furthermore, such a manmade construction masquerading behind the metaphysical, or obscured by a noumenal world, will be open to manipulation by the main beneficiaries of such a fabrication—with perpetual power-struggles over protecting the delusions of competing views. Such struggles have historically caused much conflict and suffering in the battle to define morality.

In order to avoid attachment to what may then also become no more than dogma under naturalism or scientism, we need to realize that any normative will unavoidably be based on what we interpret as how it *ought* to be, a *temporary* temporal 'what is' in what is referred to in this book as our **Physical sphere of reasoning** (PSR)—this term and its relatives (**Logical sphere** and **Metaphysical**) to be defined and explained in the next chapter and throughout the remainder of the text. In other words, our assumptions of how the universe works or what is ethical is based on a rather fragile 'how we think it ought to be' as evolving organic lifeforms, attempting to survive while continuously formulating transmutable ideas. We in turn continuously circulate these ideas between physical realities (Physical sphere), uncertainties (Logical sphere), and the unknown. It should be obvious without much discourse that without constant pliable interchanges between 'what is' and what we think 'ought to be' we cannot evolve a truthful epistemology of temporary acceptable 'what-is's or any realistic theories. This presents us with a morality that is in the same position, where we base our 'ought-to-be's (normative) on 'what-is's that used to be 'what-ought-to-be's and subsisted progressive rational criticism. We unavoidably always have to return to face the *what 'is'* in our Physical sphere of reasoning in an interconnected evolving universe that simultaneously prescribes and describes in an interactive and interconnected constantly *changing* system. This interconnected system continuously evolves and enhances itself by exchanging ideas within a principled perceptive network, in a critical, rational, and 'falsifiable' manner.

Karl Popper, perhaps the most famous science philosopher of the twentieth century, proposed the idea of **falsification**. In simple terms, falsification is the methodology whereby science derives answers by a process of refutations of hypotheses that can be proven false rather than authentications of what is true. By eliminating all the false hypotheses with certainty, one gains confidence in the validity of that which remains unrefuted. Popper urges us, at a minimum, to pay more attention to the uncertainties and biases in our thinking that may turn a truthful science into a misdirected pseudoscience such as we often find in today's marketplace, as clarity and truth then become heavily afflicted by the profit motive.

* * *

The next question we inevitably confront is how a creature like ourselves or objects that cannot representationally recognize anything can have evolutionary origins without valuing anything? Fortunately, recent revelations in genetics and neuroscience are setting new guidelines. DNA is revealing itself as a recognition system. Basically, recognition involves valuing: to sense it and want to interact, or want to get away. Valuing takes place when it can be *interconnected* with everything else in order to create a pliable valuation *system* and formulate workable operations, which we can think of as 'ideas,' like the idea to build a protein. At the very least the first strands of RNA had to recognize or identify (biochemically through receptor sites) the presence and *value* of transcriptase enzymes on the physiological level, interconnected to an environment that gave it the *idea* to replicate itself in a *changeable* manner. Backed not only by new evidence in science but by using orthodox logic (deductive and inductive), the whole principle of evolution is now seen as based on an interconnected network of pliable *recognition* systems, on all levels. Recognition and interaction occur on various levels from atoms to DNA, escalating into complex organic life. Evolution cannot operate in isolation, and in order to make contact it needs to be perceptive. So, we can safely have an impression that there are universal obligations and workable rules to interconnect and get closer to temporary workable 'ideas' within a pliable and progressive recognition system—whether on the molecular, cellular, or social level. The temporary values proposed here as pragmatically advancing our Physical sphere of reasoning (PSR), as we shall define and discuss in chapters 1 and 2, are purposely driven to expand a *progressive interconnected and escalating perceptive network.* Circulating and valuing ideas for pragmatic value between the Logical (uncertainty) and metaphysical (unknowns) spheres, such a network is dependable on reliable interchanges. This new understanding, mimicking what has recently been witnessed in biology, reveals a system generating complexity as it expands its interconnections and evolves its perceptive mechanisms. Discrediting an era of focused reductionism and the limitations set on measuring matter, this new perception will also act as a release from the strife created by competing uncertainties in our Logical sphere and help to reduce the doubts in our Physical sphere that accompany all our thoughts, enabling us to take advantage of our full creative potential.

* * *

We can now consider ethics to be *a guide for an interconnected interdependent group; a guide that respects the joint origins of the group and their united concerns as linked throughout a shared environment.* We can now evolve what the guide will say, as well as the definitions of all the terms, as we better understand the interplay between our genetics and a dynamic environment. In other words, this definition allows for both co-evolving values and pragmatic ideas, and a progression of our *moral* demands on ourselves and one another. RNA and DNA coding is such a system in its elementary form. The brain and its primitive precursors are interconnected products of a similar recognition system.

Once recognition takes place, we can judge (value) and then interconnect. Changing ideas can emanate from this process and be tested against the experience of the evolutionary drive in a falsifiable manner—constantly refining our search for a better world. The definition of 'better world' in turn simultaneously changes, intricately connected to an evolving perception of these changes. As we now know, this process mirrors that of the pliable and mobile DNA, as will be discussed in more detail in a later section. We can perhaps claim that there can be no fixed values, unbendable genetic blueprints, unfalsifiable theories, or concrete ideas in a perceptive evolution where everything is interdependent and based on progressive experience and knowledge. This interdependency is also what advances this perception in complexity.

Even then, at the very best, this system will only provide temporary values about value; or 'what-is's and 'what-ought-to-be's in a state of constant change and correction. The former claims will be explained and discussed in following chapters as essential for any progressive evolution or pragmatic ethic to sustain itself. Likewise, genetic coding (life) cannot be based on anti-realism or reductionism. And neither can it be void of some form of adaptable conduct (ethic) and subject only to a natural selection driven by a static blueprint. Gaining support of the genetic code acting not only as a moral code but interconnected to an evolutionary cognition now becomes a prerequisite for making sense of an expansive evolution. Acknowledgment of such a code is imperative for our future evolutionary success and shelters us from the damaging effects of external dogma, reductionism, and false belief systems misdirecting our inevitable universal evolutionary co-morality.

The so-called Darwinian dilemma (trying to detach and elevate moral realism from an interconnected natural selection) based on our argument now also becomes the Darwinian reality. We simply cannot relate to an evolution without a valuation system exposed to constant change, as part of an interconnected ongoing progressive recognition system behaving 'morally' on all levels of its network. This should be interpreted as much more than occasional mechanical adaptations or freak mutations befitting environmental demands, but as an active continuous perceptive transformation. We now see it as an amalgamation of valuation systems, functioning on various levels, from atoms to cells, organs, and organisms. Each of these immensely interconnected 'perceptive mechanisms' (regardless of it being an atom, cell, or higher mammal) operates within a network of 'ethical' demands. We shall at times refer to the objective individual (regardless of who or what) as *a* in this text—with *a* constantly formulating values and ideas about *b* while *b* is concurrently valuing, interacting, and formulating ideas about *a*. All as part of a complex network. No idea or concept can ever be more than a temporary idea of an 'experienced *a*' about *b*, or an experienced *b* about *a*—synchronously entrapped in continuous and evolving change. This constant interaction between *a* and *b*, regardless of whether *b* is change in an environment or another person or object, while all are simultaneously evolving, is not only vital to drive evolution, but also, as a perceptive living network, is our only protection against a fraudulent epistemology—the integrity of this network the key to our ongoing survival. It is here, with such delicate interactions, that our morality is persistently co-evolving with our perception, *both* internally *and* externally.

We now have a more equitable, complete, and updated version of evolution. Evolution can now be known as a highly interconnected *perceptive living-system*, following progressive principled rules. Seen as a pliable flow of 'ideas and values' collaborating with constantly changing environments, it is a continually changing set of ideas about ideas (or values about value) that adjust to a group's interconnected concerns. Such a more considerate and collaborative evolution is not only more comprehensive and more adaptable, but also simultaneously *re-invents* itself as it evolves in both intelligence and complexity as a progressive living network.

Another flaw of the old model of evolution was the emphasis on 'Darwinian success,' a goal measured by reproductive successes that was key to the survival of the fittest. Now updated, production and survival are seen as mere methods employed to continue the propagation of innovative ideas in an interconnected perceptive network, with genes and organisms as implements, not ends in themselves. The emphasis has shifted. A new evolution is revealed as goal-directed in advancing a progressive perceptive network, rather than the reproduction of specific bits of genetic matter, fighting with each other for survival. Clearly this paradigm shift also places more emphasis on *coexistence* and renewed focus on better understanding these principled interactions and their operations within their networks. On all levels more is needed to explain how complexity appears to simultaneously evolve. We need to grasp what evolution still has to teach us, so we too can successfully evolve.

The three spheres of reasoning introduced in this book represent a practical new way of thinking about reasoning. It is a method that will clarify how we perceive reality, and thus help us achieve humanity's potential. There are three main qualities of the human brain we can enhance by employing three spheres of reasoning:

1. Pragmatic thinking—so that our creative ideas can better achieve the results we intend
2. Resilience against manipulation—so that we will be less vulner-

able to advertising, pseudoscience, and rigid reductionism
3. Greater adaptability and pliancy—so that our minds will better adapt to changing conditions and better incorporate new information into our understanding of reality.

Together these abilities can help us to avoid getting stuck in old thinking or blocked from finding a clear forward path. These three spheres interact in unison as the Physical sphere of reasoning (PSR), Logical sphere of reasoning (LSR), and the abstract yet vital Metaphysical sphere (MS).

* * *

The *Physical sphere of reasoning* (PSR) is where we contain the verifiable, workable ideas about our physical world. This is where an empirical science mostly operates, for example by finding and eliminating errors. It is a sphere of physical realities, functional theories, and applicable mathematical equations. It delivers pragmatic results: rockets that send humans to the moon, surgeries that heal, skyscrapers, the Internet. An important feature of the Physical sphere of reasoning is its dependency on what, who, when, and where you are. Imagine the Physical sphere of Columbus compared to that of the average human today. The security provided by the high bar of entry into this sphere also makes it difficult (though not impossible) for false ideas or manipulated evidence to creep in. Yet another important feature is that the ideas in this sphere are constantly being adjusted and adapted as new knowledge enters it. Nothing is permanent. For example, Newton's laws of physics had to adapt to the arrival of quantum mechanics. The virtue of this lack of permanence is that it allows for progress—the evolution of knowledge. In this way, the PSR mirrors the principles of evolutionary biology, as our genes themselves evolve from generation to generation. Guided by both internal and external principles, the PSR constantly interacts with the much less certain Logical and Metaphysical spheres.

* * *

The *Logical sphere of reasoning* (LSR) is the sphere where ideas are considered and evaluated to determine whether or not they can be placed within the Physical sphere of reasoning. This sphere is where hypotheses are tested. Although full of uncertainty and doubt, this sphere relies on sound logic, scientific methodology, and reliable perceptions in order to arrive at valid conclusions. Due to the uncertainty of the concepts being evaluated in this sphere, there is always the possibility that personal biases, deliberate manipulation, or simple lack of information might lead us into error. Therefore we have to be careful to only hold *tentatively* any ideas that are in this space. Here's a simple example: If we see an apple on the table, we can pick it up, taste it even, and thus verify it is an apple. So, it belongs in the Physical sphere. But if we imagine there might be an apple waiting for us on our desk at work, where we left it last night, we can't know it for certain (someone may have eaten it). So that idea of an apple that we only contemplate rests in the Logical sphere. Other ideas in the Logical sphere include the possibility of microbial life on other planets, or the health benefits of certain traditional medicines that have not been rigorously tested. It also includes new ideas in subatomic physics, predictions about the stock market, and most other economic predictions. And it would include unscientific but still potentially testable ideas—conspiracy theories, the existence of fairies, heaven and hell, even God.

We can perhaps see why our current thinking is often in turmoil: because we fail to clearly distinguish between ideas that have been validated, and those we hold due to belief. When our beliefs conflict with the facts about reality, all too often we choose the familiar over the rational. The desire to impose our personalized beliefs on others has been the cause of much human conflict and suffering. But there could be great social utility to this distinction between spheres: That which is in the Physical sphere has been thoroughly tested, and so those ideas become a common ground that people can agree is true. Clarifying difference between the spheres can give each person a better possibility of recognizing an error of reasoning, and adjusting their understanding to better fit the facts.

* * *

The *Metaphysical sphere* can also be called the sphere of the unknown. It includes the whole great realm of existence that humanity has not yet contemplated, explored, or discovered. *Meta-physical* literally means 'beyond the physical,' and so we can take it to mean that which is beyond the Physical sphere of reasoning. In this sphere lie our future discoveries and unthought-of experiences. Since we have not yet *thought* what is in this sphere, we can't really call it a sphere of *reason*. Yet the ideas we generate about the Metaphysical are vital to humanity. As George Bernard Shaw wrote: "Some men see things as they are and say, why? I dream things that never were and say, why not?" By contemplating the Metaphysical, we create, invent, imagine, and conceive new thoughts.

One advantage of including the Metaphysical as a sphere is to escape from reductionism. A reductionist mindset asserts that only the physical exists. If something can't be tested and validated by science, it can't be considered real. But such an approach can easily also be responsible for clouding our imagination and hampering progress. Just as we are developing a new understanding of a perceptive evolution, so too our own ideas about reality are constantly evolving. The Metaphysical forms the substrate from which these new ideas come into our minds. It gives us mental material with which to think new thoughts and stimulates our mental evolution—most vitally in response to new challenges in our human environment. In this way, the Metaphysical can be seen, like the other two spheres, as an actively evolving sphere, wherein the previously unperceived is turned into the newly perceived. There are endless unknown answers, unimagined and unformulated questions out there in this Metaphysical sphere of our existence. This vital sphere interconnects and interacts with the other two spheres, and we simply cannot evolve a sound epistemology without it.

To further clarify: The idea of a Higgs boson particle used to belong in the Metaphysical sphere (MS)—an unthought-of unknown, until Peter Higgs thought of it. As soon as he thought of it, the new idea shifted to the Logical sphere of reasoning (LSR), where it was studied, debated, and tested. As soon as it was validated, the idea moved to the Physical sphere of reasoning (PSR), where it is now taught in physics classes as a fact about the world. However, as the realm of subatomic physics so well demonstrates, just because an idea is in the Physical sphere today doesn't mean the idea might not be sent back to the LSR tomorrow as new evidence comes into our awareness—or entire new paradigms. So even the Physical space of our knowledge remains unfixed, and constantly adjusts to new information. This is essential if it—and we—are to continue to evolve and adapt. The significant adjustments that scientists now witness in our genome, interlinked and continuously interacting with our epigenome and environment, also reemphasize the importance of the unimagined in a rapidly evolving world.

* * *

It is suitable to ask here: *Can we accept the progressive concept of a perceptively motivated evolution presented in this proposal?* We can use a diagram to illustrate this (see Figure 1).

All three spheres continuously interact with, stimulate, and advance each other as they evolve through time, in unison. The interconnections between these spheres will be illustrated in subsequent chapters. This system serves as a *method* to evolve in complexity as it circulates and advances concepts. Directed to continuously evolve better ideas in an interconnected network, this concept is similar to how a perceptive evolution based on new understanding adapts through time.

The urgent value of a sober, candid, and *ethically* driven yet pliable Physical sphere of reasoning is imperative to avoid unnecessary complexities and relativities in such a changing and adaptable network. To confront the challenging task of evolving pragmatic and reliable concepts in this changing intricately interconnected world, this sphere

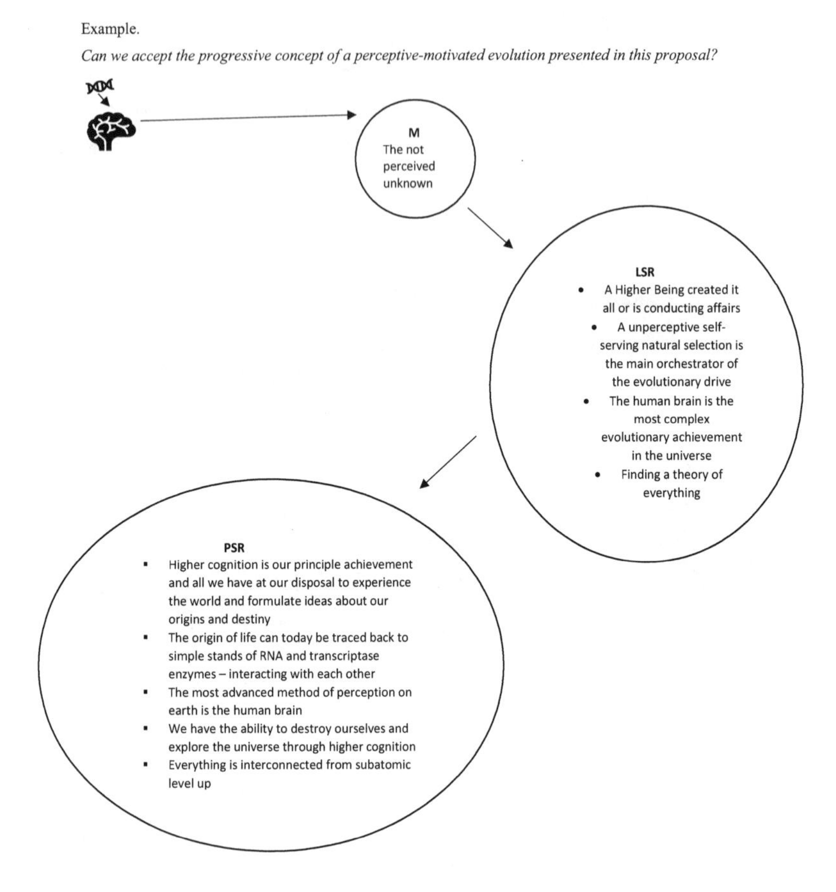


**Figure 1**

has to be simultaneously insightful and adaptable.

The need for truth and pragmatism in a Physical sphere is also especially important in a future healthcare system where our responsibility now extends, based on emerging new evidence, to the pliable and transgenerational passage of the genome. This moral duty of our Physical sphere of reasoning to evolve interchange workable ideas in creating a better and safer world extends to future generations. The Physical sphere would anchor those ideas that create a better, safer, and healthier world for us *all*. Those that do not—the fraudulent, the manipulative, the pseudoscientific ideas—these would be consigned to the Logical sphere, or else returned to the Metaphysical.

Through interchanges between the Metaphysical, Physical, and Logical spheres of reasoning we can discover, debate, and secure an adaptable ethic, and thus prevent an external moralism and reductionism from setting barriers to our progress. In developing pragmatic but open systems and applying them to our society—healthcare, education, economics, politics—we can perhaps evolve as a more benevolent, moral, and perceptive society, as we are designed to do. This principled evolution is more goal-directed as a united goal than an egotistical drive to reproduce and survive. We have now also entered a new phase in our interconnected evolutionary advancement with a complex genepool full of potential ideas at our disposal, where egocentricism and greed will be exposed as having little utilitarian or genetic worth. In fact they have become the greatest threats to a truthful science and the search for a better future for humanity, its freedom and life on earth.

## 1.2 Logic

We must acknowledge for purposes here the enormous discourse that topics such as reality, existence, and logic can trigger in philosophical circles. The complexity and importance of logic, and the interest that Artificial Intelligence (AI) and robotics have recently triggered, should also not be held in disregard. To create a distinct demarcation between an Artificial Intelligence and organic cognition, I adopted a naturalistic approach where pragmatic logical outcomes are reflected in our Physical sphere of reasoning (PSR) in constant interchange with the other two spheres, the LSR and MS.

Logic, semantically put, would be most likely seen as temporal for our purposes here. Let us consider Arthur Prior's **tense logic** where:

P "It has at some time been the case that..."
F "It will at some time be the case that..."
H "It has always been the case that..."
G "It will always be the case that..."

We cannot include *tense* in the temporal logic suggested in my argument. This is in principle because all of cognition, ideas, and knowledge is in a constant state of flux and simultaneously interconnected to change, the past, present, and now. I see tense logic as reductionist and in conflict with any proposal as an attempt to fix knowledge and cognition (either present, past, or future) in time as a constant non-changeable extant, which is, as is explained, not possible in evolution. In our proposed model of Metaphysical sphere, Physical sphere of reasoning, and Logical sphere of reasoning constantly interacting and exchanging ideas, knowledge and cognition are not only exempt from this conflict but constantly evolve in synchrony with change and time. This fits our new understanding of a non-reductionist evolution.

Our logical and naturalistic approach of an interconnected, malleable, and percipient Physical sphere will also reduce the semantic complications that logic, tense, and existence inevitably will always fall victim to. I also see knowledge acquirement in scientific research certainly and inevitably as not being free of induction logic (evidential support) and falsification but in my proposal fully in acknowledgment of the proposal by Moritz Schlick: '*always reserved and temporary in light of further experience*' (*Die Naturwissenschaften* 19, p156 [1931]). It should be therefore treated cautiously and without dogma, scientific or other, and yet find security in a pliable Physical sphere, focused on ethical advancement of an authentic evolutionary cognition.

Existence is seen from a pragmatic, naturalistic, and 'medical scientific' viewpoint, where objective values determine whether an organism is alive or dead as a perceptive being, but carried further to how its actions are conducive to an improved quality of life, harmonized as an interconnected concern. The latter is a prerequisite for a life in the traditional sense, consisting of interconnected projects, concerns, and relationships. The subjectivity of quality of life (QOL), of being alive without having a life (in a vegetative state), and the more recent intricacy of Artificial Intelligence are well acknowledged, with complexities seen as part of a pliable, perceptive whole, evolving both morality and cognition in synchrony. Any human creation will remain utilitarian in such a complex network that stretches transgenerationally for eons, driven by an *immeasurable intelligence.*

* * *

The traditional philosophical four notions of logic, L1 to L4, are simplified but not ignored. We employ evolutionary concepts of cognition in support of all four valid inferences (L1) and logical consequences (L2 logic), logical truths (L3), and the form of judgments (L4). This is done in an interchange between our Logical and our Physical spheres of reasoning to avoid the semantics that may lead to criticism of an entirely philosophical, or the condemnation of a naturalistic, concept of logic and reality. Cognition is taken as idea-making, tested against constantly changing environments. My argument is again here strongly supported by the recently accepted active anatomical and physiological *plasticity* of an evolutionary cognition set in a *mobile* DNA. This remains the primary and only known means with which to be cognitive of a Physical sphere of reasoning and experience the world and universe unfolding around us. Inescapably, as organisms are changing and constantly adapting (immensely interconnected and interdependent) as part of an active evolution, a fixed objective world, following the same basic rules and 'true' judgments, tense, or logic persistently, now becomes absurd. Imagining that such a complex evolving network might lack perception becomes impossible.

* * *

As mentioned, we should at all costs prevent a pseudoscience or erroneous beliefs from driving our evolving epistemology. Such a demarcation is vital, not only in a Popperian sense where 'testifiability and falsification' (a truthful flawed idea) is a necessity for progressive knowledge and where science remains the most important key; but also, where a pliable ethic free of dogma and manipulation can help us to protect knowledge from suppression, retention, egocentrism, and short-sightedness. Safeguarding against such manipulation is not only essential for a progressive and sagacious cognitive evolution and the advancement of sound knowledge, but is also our duty in leaving behind a praiseworthy and pragmatic morality in a functional world for future generations to follow.

It is witnessed throughout history that the deleterious effects of our moral inadequacies will have lasting effects for generations to come, should we fail to match our evolutionary demands. Some scientists are now also witnessing and warning us about the harmful effects of some of the chemicals we so easily distribute, such as pesticides and petrochemicals, not just on the environment (in a growing and still unexplored list) but also on genomes, this damage evident in rats extending transgenerationally for up to ten generations.

## 1.3 Key Areas of Concern

### *Healthcare*

Practitioners of medical science, as caretakers of our physical and mental health, develop and evaluate treatment outcomes according to scientific technique and evidence in our Physical sphere of reasoning. I acknowledge the enormous debate that topics such as evidence-based studies, placebo effects, cultural beliefs, reality, and logic can evoke in healthcare outcomes. Growing pseudoscientific activities, mostly fueled by financial gain—with 'pharma' standing out here—also give rise to ineffective and sometimes dangerous interventions with the growing use of supplementary, unnecessary, or 'natural' remedies. Mostly with minimal, if any, impact on healthcare outcomes and operating in our Logical sphere, the environmental and long-term genomic impact of concentrating these chemical structures is still vastly overlooked, poorly researched, or ignored. The financially driven pharmaceutical industry is also not helping to boost an ethic in current healthcare with its increased focus on market trends and marketing techniques, principally driven by huge profits. Healthcare providers, insurers, government authorities, and most importantly patients, need guidance and clarity on how to distinguish between medical science and medical pseudoscience and solipsistic misdirected 'profit-seeking only' interests in healthcare.

This need was highlighted by many; among the more recent studies is work done by Woodhandler and Himmelstein (*BMJ* 345, pp50–1 [2012]). Here they express legitimate concern that offering financial incentives may negatively impact on the more noble motivation of healthcare—clinical excellence and altruism. Arnold Eiser, in *The Ethos of Medicine in Postmodern America* (2014), recorded bias in 69 percent of called-on experts in their fields when acting in advisory capacities based on surveys. A disturbing 75 percent or more of clinical research published in leading journals is today funded by pharmaceutical companies and medical device manufacturers where profit is mostly the major, if not only drive. We are already aware that pharmaceutical companies place profit maximization over scientific objectivity. In much of the research today the main incentive is now marketability and profitability, and with new drug development mostly funded by pharmaceutical companies the goal and bias lie in proving efficacy of a drug rather than possible shortcomings—manipulating a Logical sphere into a Physical sphere. Negative support may either be ignored, understated, or not published. Based on recent evidence in the United States, amalgamation and inevitably the monopolization that follows by the financial powers behind this drive have also cost the consumer of healthcare increasingly more and funneled wealth to make a select few in the corporate world extremely rich. Officials are also easily bribed under such a system to clear distribution of novelty drugs and uncertain techniques that save few lives but claim to affect healthcare outcomes or fund future research with narrow margins. In veterinary medicine, this more callous corporate mentality, with its focus on financial reward instead of on clinical excellence and a pragmatic, ethically applied universal and multidisciplinary healthcare, has also infiltrated deeply on all levels in recent decades. Last and not least, as stated by Eiser, "the corporate mentality, promising savings and improved standards, in healthcare has merely increased neglect of experiential and cultural aspects of healthcare and moved it from a more logical anthropological model to a callous business one" (A. Eiser, *The Ethos of Medicine in Postmodern America*, 2014, p5). From a public perspective, the practitioner is now clearly a subordinate part of the corporate world.

### *Judicial*

It is becoming more difficult for courts to get the facts right in such confusing times, and easier to bribe corrupt officials. The reliability and sources of diverse types of evidence presented to courts, seen as correctly determined evidence based on expert testimony and candid knowledge, is becoming complex and debatable—even among called-on experts. Sometimes it is in the interest of litigants to present non-scientific claims as solid science backed by some sort of research and peer-reviewed publication dug up somewhere. Therefore, courts and ethics committees must also be able to distinguish between science and pseudoscience. A universal ethic among healthcare workers and pharmaceutical companies will be welcomed by all as a growing number of issues concerning healthcare are based on legal, political, and financial muscle, with the stakeholders driven by profits in a trillion-dollar industry, rather than confronting sincere practice-based patient and moral concerns. R.G. Steen (2011) observed that the 742 English-language research papers retracted from the PubMed database between 2000 and 2010 had error or misconduct (73.8 percent) and fraud (26.6 percent) as reasons for retraction of papers. Other researchers, such as Felicitas Hesselmann et al. (*Current Sociology Review* 65(6), pp814–84 [2017]), pointed to the fact that although the extraction process is helpful to the scientific process, "its principal value is the fact that it creates awareness that misconduct exists. As a consequence, attention is mainly drawn to the fact that misconduct exists and that someone is dealing with it in the interest of the scientific community; who this is and how they are doing this, remains opaque."

### *Environmental policies*

To be on the safe side against impending natural disasters, it may be legitimate to take preventive measures when there is valid yet insufficient evidence of an environmental hazard. This must be distinguished from taking measures against an alleged hazard for which there is no valid evidence at all or misdirecting areas of major concern due to personal political or commercial interests. Therefore, decision-makers in environmental policy must be able to distinguish between scientific and pseudoscientific claims in research if they aim to have a realistic and pragmatic outcome. We also urgently need a more universally adaptable system and ethic here to measure the impact on life on earth, and as already mentioned, now also the *genome* as part of such a living network.

### *Science in education*

The promoters of some pseudoscience (notably creationism, financial enterprises and alternative remedies) continuously and increasingly try to introduce their teachings and views in school and university curricula backed by growing financial support. Teachers and educational authorities need to have clear criteria of inclusion that protect students against unreliable and disproved marketing strategies by powerful financial institutions, stagnating a more pragmatic knowledge. Commercial interests and support may furthermore also swing education to take a turn often different from logic or objective fact but influenced by personal religious or sociopolitical structures.

### *Cultural*

With a growing number of professionals moving freely between countries and in general a more mobile global workforce, awareness of cultural differences has become more pressing (albeit simultaneously becoming more uniform) as a factor for healthcare workers to consider in their care. There is a slowly emerging, more progressive universal knowledge-base in healthcare, but it is still very vulnerable to being hijacked and manipulated by major commercial interests and big corporations grounded in their own set views and interests. Besides the biases of overpowering corporations and pharmaceutical companies, clinics promoting alternative remedies lacking sufficient evidence and simultaneously practicing medicine as a science have become more common. Such 'holistic' clinics see these cultural gaps and remedies (in need of much more substantial research) often more as a cultural belief system causing no harm with minimal impact but with added financial benefit to the facility in competitive markets. There is also now, to top it off, growing concern, expressed in numerous recent publications, about the impact on the mental health and ethics of doctors under corporate control. Besides affecting the culture and ethos of healthcare application, concern should also be raised about the use of certain expensive medicines, without a significant or clear positive impact on overall healthcare outcomes in patients already burdened by financial constraints. Often biases and politics can affect the distribution and promotion of such defeasible medicines.

With the inarguable acceptance and realization of the interconnection between evolutionary aspects of cognition, knowledge and culture as a global but vulnerable cognition-gaining process affecting us all, as suggested by Campbell (1974); Lorenz (1977); Riedl (1984); Wuketits (1986), we can see the enormous and delicate moral responsibility we all now carry as, unavoidably, a *global unit*. Such obligations and responsibility are in urgent need of updating to meet universally acceptable progressive new moral demands.

* * *

From such understanding, the following three essential key areas emerge, then, in a search for a pragmatic and universal morality.

**1) Evolutionary cognition**—defined as an adaptable mental action or process of acquiring knowledge and understanding of a changing world through thought (ideas and genetic adjustments), experience, and the senses; this involves the continuous evolutionary trial-and-error application of 'ideas' tested against the objective realities that an organism encounters in its physical world. Such *idea testing* can be broken down to the biological, physiological, biochemical, and atomic level and is conducted and best understood as part of a burgeoning interconnected network, evolving and operating under a sound universal ethic.

**2) Knowledge**—defined as facts, information, and skills acquired through experience as part of a progressive perceptive evolutionary process. Tested ideas, compliantly established in our Physical sphere of reasoning, improve our quality of life, understanding, and chances of survival within and as part of an interconnected evolving bio-unit—not merely as an aid in a struggle to survive or outcompete one another in isolated groups.

It is *only once we accept and progress to see our world as an evolving non-fraudulent, highly interconnected system, in an expanding epistemology with an inbuilt tolerance and morality, that we can apply truthful knowledge and live ethical fulfilling lives without anger or fear.*

**3) Culture**—defined as intertwined with our evolving knowledge, ideas, and belief systems (religions) such as the arts and other manifestations of human intellectual achievement regarded collectively. It is regarded as interconnected with our evolutionary epistemology and belief systems and with the potential to slow progressive knowledge when subject to restrictive hermeneutics or dogma but also the potential to extract new ideas from the Metaphysical and Logical spheres. This exists rather abstractly but as essential 'ideas about ideas' axiomatically in our Logical sphere of reasoning. Culture and belief systems have a marked impact on ethical behavior and an evolving epistemology.

## 1.4 Approach

### *Cognition*

Inarguably, the most important part of our existence as humans is our ability to perceive. It is the only means by which we can gather knowledge about our world, formulate and relay our thoughts, and create ideas, art, or music. A biological epigenetic understanding of cognizance, being the key to any awareness, must therefore be included in any realistic and practical understanding of progressive ethics today. This cognition is now evident as products of an interconnected mind-body function in response to and interacting with a simultaneously co-evolving habitat, as a pliable living network.

In the neurosciences today, we can also discuss the issue of the specious present as dependent on the organism, the stimulus, and its physiology. Time is viewed as an individualized evolutionarily (physiologically) decided concept with variation between organisms depending on *what*, *where*, and *when* you are there. We may be staring at the spider on the wall and both live in the same time, but our perception and concept of time will vary markedly. We have evolved our concept of time because cognizance operates in a framework of time, sphere, and presence, driven by the physiological need to obtain food and shelter and evolve ideas in a constantly changing continuity. Time is also dependent on change, and change cannot take place independently of time.

A clear and updated concept of evolutionary cognition is needed where ideas related to 'mind' have traditionally focused on the metaphysics and epistemology of mind in creatures that have language, so were centered around semantics and humans. Today, due to advances in genetics, the biological capacity for language may be more accurately described as a collection of evolutionary biological capacities, most of which we share with other species co-dependent on change, perception, and time.

Historically, researchers were also hindered by whether animals are minded or rational, and whether they have concepts or beliefs, but they have struggled with the issue of how to answer such questions given the inherent limitations of their investigations. The main reason for such bygone limitations was the lack of association and use of mainly behaviorism, the application of language, and psychology as principal tools for their research—all subjective and disjointed from the objective proof needed by science. In an era of genomics and backed by new technology, research in the epigenetics of neurodevelopment has decoded cognition and mind as an interconnected naturalistic phenomenon. In medical science and epidemiology, it can even be seen as detrimental to healthcare outcomes to not do so. The 'lived-body' notion maintains that bodies are not objects, but "multi phasic, experiential beings of living systems that have come to be seen as systems (of which mind and body are a unit) which are integral parts of larger systems, in *permanent interaction* with their environment and capable of constructing their own subjective realities" (Sprenger, 2005). Numerous publications in evolutionary biology now act in support of this concept.

With such understanding emerging in biological sciences, neuroscience, and genetics, the 'living systems,' phylogenetic aspects, mind-body unit, and interconnection of cognition now act as *a priori* support for our interconnected evolutionary links. This is further backed by new techniques in Magnetic Resonance Imaging (MRI) of active brains in different mammals and humans, showing only minor differences. Innovative technology, having mapped the genes in a growing number of species, is also showing that only subtle changes on the genome resulted in escalating perceptive abilities between species with transgenerational adaptations.

* * *

Briefly reflecting on the early history and influence of Western philosophy here, it showed a tendency to focus on the Aristotelian 'rationality' of humans and to see animals as lacking rationality and therefore, as is well known today, somewhat confused the issue of animal welfare. Aristotle defined "human" as "the rational animal," thus rejecting the possibility that any other species is rational (Aristotle, *On Metaphysics*) and so setting a Western tradition of neglect of other sentient beings in this regard. Later, St Thomas Aquinas (1225–74) followed this tradition by claiming that animals are irrational because they are not free (Aquinas, *Summa Theologica*). Centuries later, Descartes defended a distinction between humans and animals based on the belief that language is a necessary condition for mind and, based on this, concluded that animals are "soulless machines." Looking for objectivity to back his concept, he allocated this significant duty to the pineal gland as the seat of the 'soul' (Descartes, *Discourse on the Method*). John Locke agreed that animals cannot think, because words are necessary for comprehending universals (Locke, *Essay Concerning Human Understanding*, 1689). Following in this tradition, Emmanuel Kant with emphasis on his Categorical Imperative concluded that "since they cannot think about themselves, animals are not rational agents and hence they only have instrumental value" (Kant's Lectures on Ethics). Yet simultaneously, the main theme of Kant's argument was duty as the end goal in ethics, this in an era where the ox still harrowed the land that produced most of the food people consumed. Kant continued in support of his claim by equating the moral doctrine as "following principles *a priori* in pure practical reason and therefore clearly separated from the doctrines of an empirical based physical natural world." He specifically singled out anthropology; the conflict in this argument is easy to see today in an era of genomics. The lack of a developed science at the time can perhaps be offered as a crestfallen excuse for such mistaken philosophical concepts, but the resultant needless suffering in all forms, including vivisection without anesthetics, is hard to forgive. The hurt that other sentient beings had to endure since then and until recent years, however, hardly satisfies as appeasement or ongoing excuse for any ignorance today.

Fortunately, there were also early dissenters proposing different but equally subjective philosophies, sadly overshadowed as always by the dogmatic ordinance of contemporary ruling culture and fashions. Voltaire criticized Descartes' view that humans but not animals have souls and hence minds, by suggesting that there is no evidence for the claim (Voltaire, *Philosophical Dictionary*). The philosopher David Hume, considered by some the father of cognitive science, was more openly dismissive of the animal mind skeptics when he claimed:

> Next to the ridicule of denying an evident truth, is that of taking much pains to defend it; and no truth appears to me more evident than that beasts are endowed with thought and reason as well as man. The arguments are in this case so obvious, that they never but escape the most stupid and ignorant.
> Hume, *Treatise of Human Nature* (1738)

Hume's statement perhaps was more empirical, lacking enough scientific evidence at the time. However, with new proof today there can no longer be found any deficiency in his claim. Today in a post-Darwinian world with neuroscience exploding with innovative ideas backed by modern technology and discoveries in phylogenetics and epigenetics, and with enough objective evidence, it would be extremely hard to philosophically or on any other level defend non-animal-linked origins of our evolution-driven cognitive functions.

With more evidence of the biological aspect and evolution of our cognitive abilities, we may also discover and understand what is required of us and how we should behave, and what sort of an ethic is evident in nature (if any). The Quinean (1969) view that we should abandon epistemology for empirical psychology is no longer widely accepted due to recent work done in the biological and neurosciences. On the other hand, the concept of **enaction**, the manner in which a subject of perception acts out the requirements of its situation, is entirely valid to present and develop a framework from. This is in support of the concept postulated by Varela, Thompson, and Rosch, in *The Embodied Mind: Cognitive science and the human experience* (1993, p197): "much of what an organism looks like and is all about is completely *underdetermined* by the constraints of survival and reproduction. Thus adaptation (in its classical sense), problem solving, simplicity in design, assimilation, external steering, and many other explanatory notions based on considerations of parsimony, not only fade into the background but must in fact be completely reassimilated into new kinds of explanatory concepts and conceptual metaphors."

These earlier contributions suggesting that the experienced world is portrayed and determined by mutual interactions between the physiology of the organism, its sensorimotor circuit and the environment are invaluable in bridging the human experience and a more objective neuroscience.

## *Knowledge*

By the nature of what it stands for and what has already been said, knowledge is continuously evolving and has a direct effect on how we exist. Knowledge contained in our Physical sphere of reasoning is both temporal and temporary and a direct spinoff of our evolving perception in a changing environment. As an interconnected concern, this also influences how we live, quality of life, religious belief, and our political structures. We are, however, slower to change the latter two arrangements because of the dogma that secures such communal structures and the familiarity they present. The effect of greed and fear is a considerable influence and of major concern, not only in misdirecting the application of new knowledge but with the dangerous potential of setting a misguided epistemology in *repressed knowledge*. The still-existing biases in research and healthcare, and corporate influence through funding on our educational institutions today, are more responsible for knowledge displacement than most of us are perhaps aware. Previously, science philosophers such as Thomas Kuhn, *Criticism of Scientific Revolutions* (1962), also raised concerns, albeit from a different angle, but expressed concern about a science lacking transgenerational continuity as being reductionist and vulnerable to culturism.

The positive impact of beauty, altruism, and interconnection as a drive for knowledge is also overshadowed by the harmful biases seen when profits, creationism, and politics at times can overshadow reality and honesty.

Our current concern here is also that our initial interpretation of an evolutionary theory may have been responsible for establishing a culture with an unnecessary iniquitous survivalist mentality, in sub-disciplines and people in general. This causes a form of harsh egoism to appear, affecting all aspects of our society, including business and healthcare. In now discovering a more *benign interdependent* version of our evolutionary roots we are also more likely to direct and develop a more realistic, trustworthy, and pragmatic science, and subsequently moral society. This indeed sets a better platform for survival on a global level with science then securing its place as invaluable in creating a better world for all of us in a universal morality.

## *Culture*

Culture emerges with a society and its epistemology, and inevitably has a strong influence on ethical and moral behavior. Expressing itself as representative of social groups with similar beliefs or interests and acting like a social glue, it subsequently may also create division between dissimilar cultures. Culture can also be a great manipulator of current knowledge; often centered around similar beliefs to gain support for members of such a group, it can become disfranchised from more universal needs. More positively, a progressive cognition constantly in search of workable contemporary ideas can employ culture to have a significant impact on the way we think, behave and act on a larger scale—perhaps as an emerging new culture in science in an age of interconnectionism. Culture also extends into the metaphysical and into the arts (discussed in later chapters). We also must acknowledge our existence in an era where cultures inevitably are all merging into a new globally interconnected 'hi-tech' society, where we are already busy evolving a culture of smartphones, social media, and Artificial Intelligence. Culture can sway our Logical sphere of reasoning with significant influence on our Physical sphere of reasoning, this potential also affecting science.

Culture through the arts can also bring us closer to the beauty of this world and life (see chapter 4, section 3). Expressing understanding and respect for each other's suffering through the arts, culture should never be ignored as one of our more noble endeavors and evolutionary recruits of a progressive human intellect—as long as it stays true to its search in creating better understanding of each other's needs and hurt in a combined destiny.

### *The Metaphysical*

Historically, ethics and moral guidance securely belonged to the Church under God's influence. The Church, belonging to God, secured its power in the Metaphysical. Kant again on this topic in his *Critique of Pure Reason* (2:66.1–6) referred to the Metaphysical as, "A dark ocean without shore and lighthouse, on which it is all too easy to lose one's way." Traditionally defined as the philosophical inquiry of a non-empirical character into the nature of existence, by its speculative nature, the Metaphysical is now seen as *vital* to steer us clear from remaining entrapped in the narrow confines of our interpretation of the objective world. In other words, as mentioned, it helps us steer clear of reductionism by stimulating our, at times, incommodious Physical sphere of reasoning. It would therefore be unwise to exclude the concept of the Metaphysical and not see its value as a vital driving force for a progressive cognition. This *synergy between the metaphysical and physical world is critical* for progressive knowledge-development, as will be explained in a later chapter.

* * *

Any attempt at formulating a progressive theory, *or* an evolutionary epistemology, functioning in a fixed physical or objective world without allowing for change and the lure of the unknown would be impossible. Constant change and motion are essentials needed to create a continuum of life and knowledge. Evolution, as an evolving perception of continuous change, constantly challenges the unknowns. Conative and adaptable to harmonize with this process of change, we are cognitive and alive.

This can be formulated, where changes interpreted by organism *a* within its habitat *b* are perceived through various physiological means and cognition interacting and formulating ideas between *a* and *b* as,

$\Delta a \approx \Delta b$.

Changes in both *a* and *b* are continuous and, based on simple deductive logic, infinite. Change cannot be considered without time as mentioned. Without much argument needed it can also be deduced then that, once an object does not change and is fixed, it becomes finite. And when everything is obsolete, static, and fixed in a 'theory of everything,' there will be no need for a perceptive evolution to confront change or acquire awareness of, or any need for, time—and no evolution or life. It is similarly hard to imagine any atom to be inactive internally and non-reactive to surrounding atoms, and, isolated from molecules and energy forces surrounding it to change into anything more. The potential of an unchanging static and finite universe is inarguably with all the scientific knowledge available to us today, simply not possible. Neither is life, cognition, or evolution without the stimulation of constant change and interchange driven by unknowns (metaphysical) a workable concept. All of evolution and life is dependent on this infinite change and interaction, based on unknowns with some evolutionary means to perceive these ongoing changes in a *network*, growing in complexity—all changing in time.

Where *a* is the perceiver and *b* is the perceived with *a* and *b* interchangeable depending on whether you are the observer or the observed, we can see how all our actions have an impact on objects (inanimate or not) around us, and in turn them on us. Harmonizing these interactions as an interconnected network of ideas with vested interests is how we grow our perception and understanding to create a better world. This process is also how our morality evolves as our awareness increases. We can formulate this as a *moral evolution*, or *Ev(mo)* in this text so that:

$$\mathrm{Ev(mo)} = \infty\Delta \mathrm{a} \approx \infty\Delta \mathrm{b}$$

We can conclude that morality is infinite and progressive.

A vital state of exchange exists between the metaphysical and physical world as follows:

Perception of objects evolves between *a*'s and *b*'s continuously interacting and evolving in complexity as everything changes—perceptive and responsive to such change. With time and change interlinked (try again here to imagine having change without time) and our perceptive evolution dependent on change, we can see the inane aim of reductionist ideologies, set theories, and fixations. With *a*'s and *b*'s interacting and interlinking concepts in Logical spheres and Physical spheres (as revealed in the next chapter), confronting the metaphysical (unknowns), the network evolves in its own complexity. I think we can, with increasing evidence today in quantum physics, claim that Newton's laws of gravity will fail to apply to all times and places in an evolving universe.

The sum of interactions between changing *a*'s, *b*'s, and unknowns networked together is where we co-evolve our perception and morality and can be formulated as:

$$\sum \Delta C(\Delta LSR \leftrightharpoons \Delta PSR) \leftrightharpoons \Delta \text{Metaphysical}$$

*where *C=cognition*

These two principles, emphasizing change, time and uncertainty, are introductory keys to a principled interconnected evolution, obliged to follow a universal ethic (code of conduct). This is also where we evolve *perception* and knowledge as we confront the never-ending changing metaphysical with infinite potential. Here we evolve cognition and morality, both internally and externally. Given then that we will never have the complete knowledge to which we might aspire, we must always strive to act accordingly in this twilight between certainty in our Physical sphere and uncertainty in the Logical sphere—between knowing and unknowing interacting with an evolving Metaphysical. This is where our morality also comes to the rescue.

Existing as progressively evolving perceptive beings interconnected to other constantly evolving objects, the quest is still: How are we supposed to behave and interact to spend good, honorable, productive, and principled lives? And why should such matters halt us from exploiting the material world with attempts to outmaneuver each other in order to survive, reproduce, or perhaps merely for entertainment value, in such a 'dog-eat-dog' society?

This significant task is explained and simplified in the following chapters, steering clear of a dogmatic science, 'trends' in philosophy, or masquerading behind the uncertainty of deities set in the metaphysical. In this task, I endeavored to adhere to the principles of an open truthful science with the aim to still create scope for continuous and attentive discussions to follow while working within a flexible model. This pliability and avoidance of reductionism, I hope, is also where we will always secure our intellectual superiority over Artificial Intelligence as we evolve in perceptive complexity.

I asked Siri what the purpose of life is. Siri replied, "I cannot answer such a question…Ha, ha, ha…" and then referred me to Google links drowned in semantics and confusion.

Maybe we can now arrive at more pragmatic deductions, while we continue to evolve in our understanding…

# 2

# An Active Perceptive Evolution—Even Darwin Would Be Surprised

*The hardest thing of all is to find a black cat in a dark room, especially if there is no cat.*
Confucius

*I fear I am losing my mind, and what would I be without it?*
Faraday (prior to his discovery of magnetic fields)

We have introduced three vital elements required to drive and explain an evolutionary cognition, with its ultimate product so far, the human brain. These are persistent *change*, the ability to *perceive*, and the impetus of the *unknown* (or uncertainty)—integral, not only to new understanding in evolutionary biology, but also to explain how the brain evolves (note tense as continuous) and functions. The latter points are unavoidable should we try to develop any realistic and pragmatic notion of the status, foundations, and scope of life, our morality, and our place in the universe. We inevitably always return to, and must ask ourselves: If the brain is not the source of developing all our current knowledge (present and past) and ideas about morality, where else can we *realistically* look?

To avoid equivocality and obtain some level of pragmatism, there is an obvious need to set clear barriers between representational ideas, the uncertain and metaphysical. To achieve this, we have introduced the value of staging these cognitive activities in the spheres of: Physical sphere of reasoning (PSR), Logical sphere of reasoning (LSR), and the Metaphysical sphere (MS). Subsequently in this section we shall explore these locutions and explain how we can circulate concepts between these spheres of perception.

## 2.1 Evolutionary Origins of Our Cognition

Light is a major carrier of information in nature. The molecular machineries translating its electromagnetic energy (photons) into the chemical language of cells transmit vital signals for adjustment of virtually every living organism from its habitat. If we take a unicellular organism such as the primeval slime mold *Neurospora crassa* as an example, it will respond to an external light source, move toward a proteinaceous broth, find 'comfort' in an interconnected group of other molds and, if conditions are suitable, replicate. Based on new evidence and thanks to new technology, we now know that transcription of genes is initiated within minutes inside cells. Here an abundance of metabolic enzymes (proteins) and their interactions are harmoniously adjusted, and subsequently, levels of certain metabolites altered to interpret and respond appropriately. (In the case of the slime mold and light, urging it to move away from the desiccating effect it may have: Schmoll and Trisch in Applied Microbiology and Biotechnology 85(5) [November 2009].)

Should such actions not be conducive to changes in the organism's environment, or should it start a novel action, a new belief or culture perhaps, such as moving into bright light or away from it too slowly, it would lead to the demise of its species and affect *others interlinked* to this infrastructure. Using this model, we can perhaps sense with some understanding of evolutionary biology how we are genetically primed to carry interconnected responsibility for our actions, even on a primeval level—and how these levels escalate as we progress. Importantly, this model extends all the way up the phylogenetic tree to eukaryotic organisms, where there is an abiding interplay between judicious decisions as an adaptable change favoring the best outcomes in a constantly changing network—this effect carried transgenerationally.

Everything is immensely interconnected and interdependent, relying not only on various physiological transmitters but also on a pattern of harmonious interactions to adjust, perceive, and aptly respond to change. In this we can already translate the origins and demands of an evolving 'ethic' (conforming conduct within a group) and moral code (correct action in support of a network) emerging as an essential basis for the continuity of life and evolving intelligence. Now also firmly backed by new evidence in the neurosciences, all these fine-tuned molecular interactions escalating in complexity, seen here and elsewhere in nature, present us with the potential to serve as a model for moral evolution.

* * *

Jumping straight into another aspect of our existence, it also recently became evident that perhaps our current form of social hierarchy set in a tenet of capitalism is non-democratic and not matching the pattern set by our evolutionary design. We see this in a growing inequality appearing in the sharing of knowledge, resources, and information. Easily manipulated by power, egocentricism, and personal beliefs, it is not matching the demands of an interactive living network with interconnected concerns, as science is now revealing to us. Isolationism, or any structure growing exclusion, will also not match the future demands facing our current healthcare system, with a new era of genomics approaching fast. With survival no longer seen as a means to an end but a mere end to the greater means, that of interconnection and progressive perception, reductionism and isolationism now appear too parsimonious to develop a progressive knowledge and evolve a reliable epistemology.

New research has dramatically changed our understanding of evolution over recent years. "Charles Darwin would be surprised" is perhaps the most apt way of introducing the impact of all these changes, as declared in the opening paragraph of the book, *Mobile DNA: Finding treasure in junk*, by Haig H. Kazazian, Jr. (2011). We can now, supported by expansive new evidence, assume that all life emerged from single strands of RNA (a molecular arrangement) orchestrated by transcriptase enzymes (protein molecules) to produce strands of protein-encoding DNA, this cascading subsequently in our ability to perceive the world as we see it. Such a reverse transcriptase enzyme has been well studied using models such as the mitochondria of the slime mold *Neurospora crassa*, exhibiting the ability to synthesize full-length DNA copies from RNA without a primer of any kind (Kuiper et al., 1990).

With the above and other recent discoveries in genetic research falsifying one of the 'Central Dogmas' of genetics where transcription of DNA into RNA was seen as the blueprint and initiator of all the diversity of lifeforms, we have to now stand back in awe again. We should now rethink the nature of our origins, with DNA standing in the shadow of RNA, and step even further back to the molecules and atoms that formed this originator RNA. This may perhaps appear as a trivial fact to some but is very significant in evolutionary biology, as we will see.

We can today continue to break down these life-generating strands of RNA and proteins to the atomic and subatomic level and here also witness their interactive behavior with transfigured understanding. We are forced to humbly and realistically accept that from such basic elements—groups of molecules, replicating strands of RNA and DNA producing an array of proteins, in response and perceptive to changes in its surroundings—originated all lifeforms known to us, including the human brain. Subsequently, so emerged our cognition and all the current knowledge we have of ourselves, our planet, and the known universe so far. The post-Darwinian struggle in accepting humankind's animal origins and interpretations of the 'image of God thesis' is now historical, and all is taken to another level with this emerging knowledge revealing an amazingly versatile, principled interconnection and interdependency between all things—not as progressive primates emerging from swamp-like creatures and amphibians. However demeaning this may still seem to some, these are impossible-to-ignore facts, progressively improving our understanding of the basic origins of our humble but escalating cognitive abilities and perhaps creating some awareness of where it is heading. We have no choice other than to pensively stand back and acknowledge this molecular origin of life and bravely start adjusting archaic beliefs and outdated ideas. Amazingly, with this also comes a much higher universal moral duty than under earlier 'set' beliefs.

We continue today in biology to break all bodily functions, perceptive mechanisms, cognition, and responses down to biochemical, physiological, and molecular-level interactions and communication in response to changes in the habitat. We operate today in an era where we have moved beyond the initial shock of realizing that higher cognition, even in its more sophisticated forms, is a result of organic evolution with subtle epigenetic adjustments over time, using biochemical means to communicate progressive change. This has now advanced to a new level in the recent mapping of the genes in a number of different species. Scientists now study a *pliable* actively changing DNA as part of an interconnected habitat, this habitat also continuously changing and being modified (by itself and us). In fact, many of the new areas of research and advances in the biosciences are slowly refocusing on this new understanding and are starting to focus on response to change on the biochemical level. Using this current knowledge and new understanding in epigenetics and phylogenetics, we simply cannot deny the minor changes on the molecular level responsible for the emergence of organisms with progressively improved and more complex perceptive abilities. In further support of this, scientists can today study and actively visualize active cognitive development by means of functional MRI, electro-micrograms, and with the aid of new staining techniques examine cellular DNA turnover in the brain. Such studies increasingly reveal the amazing and interactive constant interchanges between everything and the changes that can be stimulated by the environment, even in the ageing brain.

Recently, more surprises were unveiled in work done utilizing such methods in both human and non-human primates, and we now also know that in some regions of the brain new cells are added for many years after birth in response to environmental changes (Bernier et al., 2002), with more support also in more recent research. It is now becoming clear that cognition is not genetically limited but constantly adjusting and changing and even enhancing itself *in* the living organism. Such changes and experience are dependent on the environments the organism is exposed to. In addition, it has been demonstrated that these postnatally derived mammalian nerve cells express both neurotransmitters and also form new synapses. They constantly encode to produce protein structures to *reach out* for each other in response to changes in their environment. At a very molecular level, we can now postulate that the processes that underlie plasticity such as the neurochemical profile of the synapses, sprouting of new dendritic spines, and growth of new axons and so forth, are not only similar between species on the molecular level but no different in the developing brain than they are in the mature brain. This is important to us for two reasons. Firstly, it now appears that the *environmental impact is much more consequential* on a *pliable cognitive development* than previously thought, and secondly, cognition in turn is an *ongoing* dynamic interconnected process simultaneously affected by and affecting the environment—the more it changes, the more it changes.

We now see cognition as an active *interconnected constantly changing* phenomenon interacting with its environment. "Environments that are enriched, harmonious and stimulating will create brain development to continue and pretty much continue to operate regardless of age" (Colcombe et al., Journal of Molecular Neuroscience 24(1), pp9–14 [2004]). Simultaneously, today humankind is *changing* the environment on never-before-seen scales.

* * *

With knowledge of both the environment and cognition in a constant state of interaction and simultaneous change, we can state, *environment ⇔ cognition*, as interactive living systems within interactive living systems. This interaction evolves a neoteric knowledge to be genetically communicated transgenerationally and influenced by slight changes in the environment. We can see similar environmental and cognitive interconnection and demand in non-mammalian species and unicellular organisms. We can revive one historically much-used example here in migratory birds to explain this. Migratory birds have always drawn much interest in such perceptive adjustability to environmental changes. These migrant birds can remember the location of a breeding ground after many months and up to a year, while the non-migratory birds can recall a breeding location for only up to 2 weeks (Mettke-Hofmann and Gwinner, 2003). This can be traced to the presence of specific genes. Although the genome is almost identical in most birds, migratory or not, now with new methods in genetic research scientists are finding subtle differences in specific locations of some genes, depending on location and routes followed. This can rather rapidly be attuned to environmental changes. Minor rearrangements in some genes have caused White Geese in Scandinavia, as an example, to breed in Lapland, adjusting their routes and breeding grounds due to heavy poaching and climate change in their traditional Scandinavia. With sedentary genes carried for many generations able to be revoked and respond to environmental changes, numerous physiological and sensory adjustments can be made, even in the same bird species, causing them to follow different flight paths.

Further support for a progressive trans-genetic, active, and pliable cognition was obtained in recently discovering that higher cognitive function in the mammalian brain continues to develop and improve with age, providing a mechanism for attending to relevant information by simultaneously inhibiting irrelevant information (Casey, Durston, and Fossella, 2001). We can now refreshingly relate to our cognitive abilities and knowledge evolving while we are evolving, interlinked and heavily influenced by changing environment and information received. This newly recognized interdependence, now seen as a delicate interconnected perceptive mechanism (C, cognition) dependent on reliable and progressive knowledge, affects the health and future of the genome.

We can now re-formulate a previously seen evolution where a genetic blueprint is hoping for a few lucky or fit adaptations to outsmart a harsh world, $\mathrm{Evo}=\Delta a(\mathrm{C})\rightleftharpoons\Delta b$, into a new less egocentric and more moral evolution, $\Delta\mathrm{C}=(\Delta a\rightleftharpoons\Delta b)$, and continue to expand on this less reductionist perspective of our origins set in a progressive perceptive network.

Cognition is now seen as an interplay and interdependency between an organism (made up of cells and molecules), organisms, and environment, fine-tuned by genes (themselves molecules) on all levels—molecular to organism to environment. This new understanding of carefully orchestrated yet pliable interdependent systems at work, intertwined within a universal perceptive drive, has made it difficult to accommodate a solipsistic survival-of-the-fittest, greed-based partisanship standing in isolation. The essential need for universal interconnected moral commitments, where egalitarianism is a united global concern in a perceptive evolution, is now of primary importance to us and our future morality.

## 2.2 Moral Essentials Revealed by Developmental Cognition

To further support our new predicate, it is essential to briefly reflect on our current knowledge of the growth and development of axons and dendrites. There is no need for a detailed account or in-depth understanding of neurology beyond the embryonic deployment of neural tubes from the ectoderm in a developing fetus for our purposes here. Neither do we need to reemphasize the similarity of this process in different species. We can pick up where the neuroderm (where embryonic neuron cells stem from) has formed and neurons start migrating to specific zones in the developing brain. The neuroderm is the outermost of the three primary germ layers of an embryo, from which the epidermis, nervous tissue, and, in vertebrates, sense organs develop. Conducted by certain genetic codes, strands of RNA produce specific proteins in an analogous manner across species to initiate brain development. Once these neurons have migrated in the embryonic brain to their specified locations, one of two things may happen. Some may form axons or dendrites and *interconnect* by means of these dendrites and axons, and others undergo programmed cell death (apoptosis) and die; a noteworthy 40–55 percent of the newly formed neurons die. The whole process is biochemically conducted and orchestrated by DNA, a branching-out and interconnecting of perceptive cells (neurons). A solitary neuron is senseless, but with mass interconnections it results in the human brain.

What is significant here is, firstly, the essential need for programmed existence failure; why the troublesome genetic expression and tedious migration of neurons to simply die off, we may ask? Throughout nature and in genetics we witness how losses can be intense but always in support of improved coexistence in an expanding network. We can syllogistically conclude that the subjugated neurons are essential to contribute in obtaining a goal and that by reducing the numbers of subjugates (or not supporting them) we will also directly reduce the numbers of established neurons and their potential to respond to diversity in their habitat—and the progressive complexity of the network.

Secondly, it is important to *not* see the subjugate neurons as unfit but rather as essential co-workers in *support* of developing a higher cognition. With vested interests, they consequently also have to be carefully nurtured for a successful cognition to continue to develop. The better tended the support neurons, the more successes we get as a result. In the end the one who obtains the medal must humbly stand back and acknowledge its position, merely as interlinked in a united effort.

If we now follow the surviving neurons (not survival of the adept but as products of others' support and input) we see, thirdly, that their main drive is *interconnection*, and they achieve this by actively sampling their environment for molecular cues. These budding axons in the developing brain are ushered by a growth cone at the tip of the growing axon to either move away or toward a target. In turn, dendrite sprouting is a result of the recently isolated CREST gene conducting proteins to form small buds in the wall of the neuron and develop into dendrites. Note, all these are protein structures of varying complexity manufactured by strands of RNA, instructed by an age-old genetic code with *pliability* in its DNA to confront the unknown. It is ultimately these new interconnections that ensure our cognitive abilities, and then also the lack of interconnection and pliability that creates loss of cognitive function and mental disease (fixations). The better the interconnections and more harmonized the interaction between neurons, and the more there are, the more comprehensively can perceptions be interconnected to environment and grow in complexity. With better interconnections, support, and pliability, the result can be better general cognitive function and adaptability.

This sprouting out of dendrites can also be seen as following a rhizomatic policy structure, such patterns seen both in the brain and other aspects of nature. These formations avoid oversimplification, over-centralization (monopolization), egocentricism, and rigidity. In a nutshell, they avoid reductionism by constantly branching out, interacting, and interconnecting. Interestingly, on a slightly different level of research, a rhizomatic structure has previously been proposed as better suited for developing humane organizations as opposed to a central bureaucracy operating in isolation (Deleuze and Guattari, 1987). We can deduce from this how centralization and monopolization, clearly not hallmarked by obduracy, are perhaps counter-productive to the core principles of a branching ethically interconnected progressive evolutionary cognition.

Creation of these interconnections between neurons (a pattern seen throughout nature) is conducted by genetic coding that took millions of years of 'fine tuning.' This process is a highly principled affair, conducted and interconnected to an established, considerate, and all-regarding code. It is no longer to be taken as the result of some random egocentric mutation with aggressive means aimed at survival and to outmaneuver others in a harsh world.

Fourthly, important to note is that it is an active, interconnected and *pliable* process constantly being adjusted throughout an organism's life to improve this perceptive network—supported by its failures. *This is a struggle not to merely survive but to advance cognition within an interconnected and pliable network.*

To continue our brief encounter of brain development, these genetically established points of contact between two neurons are called **synapses** and there are two ways neurons establish contact in the developing brain through a process called **synaptogenesis**. The one is electrical (ions and electrons) and the other via more complex biochemical neurotransmitters. Even on this mechanical level there is evidence of plasticity, depending on use or disuse. Such communication links after a developmental overproduction follow the Hebbian principle where the ones not used are weakened or inactivated and those used frequently continue to function and adjust to changing needs. The system of plasticity continues in the adult brain and depends on activity once synaptic contact has been established. Interestingly, inactive synaptic contact can be triggered to function again and the Hebbian principle overcome if stimulated, even in the older individual. We shall later see the similarity in spherical idea-formation, where ideas interchange between Logical, Physical, and Metaphysical spheres of perception and can be revived from any sphere at any time, including the Metaphysical, while principally functioning in a reliable Physical sphere.

The fourth characteristic of such a rhizomatic pattern then, besides support needed in 'failures' as described above, is *pliancy.* Inevitably, an essential requirement for evolution and the continuation of life is to be perceptive to constant change. Without this quality it will fail to adapt and adjust to change. There is no place for overwhelming dogma in such an interdependent, pliant, and interconnected system, where regard for every part of the network is inarguably considered sapient and essential on all levels of existence.

Also, more recently coming to light is the previously thought mandatory value in the functions of glial support and astrocytes in this previously ignored 'senseless' brain substance. We now know that as part of this living matrix they are actively involved in myelination of some axons, causing faster-speed neurons to emerge and the astrocytes to assist in conducting signals and help repair neurons as they are needed in response to changing environmental needs, further adding to this pliancy and adaptability. This shows an unalloyed essential interdependency between need and environment, backed by what can also be called a form of 'malleable altruism.' With evidence now emerging from all over in biology of an interactive interdependency that cannot function in isolation, the more despotic reductionist view of evolution is slowly becoming obsolete. With emphasis now on these interconnections we should again be careful in our search in looking for set patterns and avoid reductionism, as perhaps in the current culture in Artificial Intelligence, with the task slowly appearing more difficult than initially assumed. We can begin to understand why—remember, perceptive pliability and change is the key to life and perception. Without change creating ongoing unpredictability there is no motive for a perceptive evolution; now we also know that evolution cannot *not* be perceptive to this change on all levels.

* * *

We may slowly also begin to detect the emergence of a new ethos in cognitive development, as we do throughout nature. Such an ethos critically depended on existence failure, support structures, interconnection, and pliancy, and we should add *trust*. With respect due on all these levels, this is no more merely seen as a tyrannical struggle between cells or organisms to dominate and out-maneuver each other in a harsh world. Such a historic and myopic view, based on self-serving 'adapt or die' principles, now appears obtuse. We now see living systems within living systems, with a unanimous motive to harmoniously interact and share the increased morality and responsibility that come with an evolved principled cognition.

Valued support for such interconnection in a constantly adaptable cognition was also gained from early philosophers, such as Donald T. Campbell (1916–96), in bringing cognitive psychology, evolutionary biology, and philosophy of science together. In creating an understanding of an evolutionary cognition and epistemology at a minimum as a combined concern and a product of humankind's social and biological progress, Campbell proposed a model of blind or **'unjustified' variation** and **selective retention** and claimed that these are necessities of any evolutionary epistemology. Perhaps we may no longer agree with

Campbell on the 'unjustified' part, with our new understanding of the transgenerational progression of knowledge, but the rest of his proposal may fit with what we now witness in both cellular communication and cognitive evolution. In the same era the philosopher Willard Van Orman Quine (1969) also suggested that we 'abandon the effort to show that we do in fact have knowledge and that we instead study the ways in which we structure beliefs and knowledge from a naturalistic point.' Seen as such, and not necessarily leaning on either Campbell or Quine for support here, the slime mold (and all life) is now seen in biology as part of an *active evolutionary living system* in a combined evolutionary cognition. This now supports the ambition of studying *how* we evolved our knowledge—*a* (observer) as an inter-acknowledged part of *b* (the observed), co-evolving a progressive epistemology.

* * *

Armed now with a refreshed understanding of interconnection, support structures, respect for failure, and plasticity on the cognitive level, even more enlightening is the recent recognition of the *mobility* of DNA in a pliant genome evinced by new research. We now know that the genome has the ability to undergo rapid changes by rearranging some of its parts as opposed to slow change afforded by nucleotide mutation rate. Thanks to science philosophers like Campbell, Riedl, and the biologist-philosopher Franz Manfred Wuketits more recently, the idea was developed of active molecular instructions *remembering* and *testing functional ideas* in an epigenetic system. Every living system is an active system and, as Popper famously proposed in 1984, "in search of a better world." With present-day understanding in biology, the latter statement of Popper is perhaps more elucidating and far reaching than was seen at the time. Based on recent discoveries, we now also know that so called 'junk' DNA, or the non-coders of the genome, are proving to be more significant than previously thought, helping to create a more flexible and active DNA—constantly 'falsifying' existing ones. Such a mobile DNA tests 'ideas' in a constantly changing habitat, perceptive as co-workers in search of a better world. It performs these cardinal duties by producing new proteins fine-tuned and sensitive to environmental challenges as an ongoing *active* process, throughout a cell's or organism's life. This may sound like a rather trivial matter to some, but the importance and impact of this in the contemporary philosophy of science is enormous.

In the context of this and with much more on offer in the scientific literature today, it can clearly be seen how an unrefined natural selection theory, supported by randomized 'lucky' mutations and the occasional despotic adaptation, is inadequate, both in current evolutionary biology and cognitive development. Clearly, such a reductionist imperative, dictated by environmental demand with cognition as a fortuitous afterthought, is not only insensitive but much too feeble to explain the complexity of the human brain and diversity of perceptive life with its adaptability and moral commitments. It is perhaps based on such a parsimonious view of Darwinism that society is also currently lacking in much needed empathy and understanding, and persistently trying to justify the inequality it forces upon itself. Such an antiquated outlook on evolution, previously seen as $\Delta a(\mathrm{C}) \leftrightharpoons \Delta b$, is also neglectful in challenging the moral and ethical commitments needed to explain a much more advanced, complex, and interconnected evolutionary biology and world, as is now evident.

The historical idea of a genetic blueprint, with natural selection reconciling a few isolated survivors with their personal demands, directed by an equally indolent environment, as the only means to host an infrequent genetic change, simply does not convince or satisfy. Similar to the active rhizomatic pattern that neurons follow in the developing brain, our growing evolutionary cognition and constant adjustments are now seen as dependent on a constant flux of interconnected ideas, ideas within ideas (habitat)—both dependent on a pensive plasticity as part of a sprouting interconnected living network. Important to us here is also how attested failed ideas are vital to advance workable ideas within a prodigious cognitive drive. We all rest heavily on the shoulders of our predecessors and their selfless contributions, mistakes, and failures, to evolve a progressive epistemology. The advent of the era of the Internet now seems inevitable in an evolving epistemology. Rather than a lucky strike of brilliance for a few to excessively draw profits from, while over half the world's population lack basic resources, we now see it as an evolutionary advent striving for interconnection that should reach everyone and help improve our morality.

We have already mentioned how 40–55 percent of neurons were needed for the development of 45–60 percent of viable neurons, the viable ones not to be established without support from the casualties, all programming the evolution of a progressive cognition. We can also syllogistically conclude that we cannot have successes without failures, and all our cognition and truth about realities progress from 'failed' ideas. This should not be seen as a cruel Darwinian struggle where only a few survivors stepped on others to emerge as victors, but as a distinct pattern aimed at improving morality and understanding, for coexistence—sifting through pragmatic and workable ideas in unison to evolve a better world. A world where workable ideas are constantly invented and re-invented from anachronistic ones to serve in a cognitive harmonious network.

We can vaguely imagine how many failed experiments, feckless (previously claimed as cures) drugs, and how many scientific papers written and exposed to constant challenges and changes have carried forward a few ideas that served humankind for lesser or longer periods. Our current acceptable ideas are bound to be replaced by better ones in the future. Using all traditional four notions of logic, L1–L4, it is easy to deduce how from a futuristic point we will all, even as winners, be considered as *invaluable failures* in time. There is no place for egoism in such a large scheme of interconnected evolving ideas. In changing our perspective on these matters, there is also no need to be threatened by Artificial Intelligence, emerging with utilitarian value in a transgenerational interconnected perceptive evolution.

With our concepts changing all the time and a pliant perceptive DNA, already designed to cope and adjust with changing demands, the technocratic era is a progressive aid in our expansive knowledge.

Technology remains as no more than a utilitarian evolutionary recruit, not to be feared. We also understand now that a vast variety of ideas are needed for evolutionary successes and progress, and that evolution stimulates itself in a perceptive network rather than in isolation.

*Any attempts made by the creation (the human brain) to modify or rig the creator (a perceptive genetic drive in a vast universe of unknowns), again using all four notions of logic, L1–L4, will merely set limitations and delay its natural design and in the end fail.* Any reductionism set in narrowly defined theories, personal world-views, or set beliefs, defending them as concrete rules, is destined to in the end also fail. We can perhaps also begin to understand that an end to knowledge is not possible because a progressive knowledge is the vital stimulant of an evolutionary cognition, and perception is vital for an evolution based on interconnection. This progressive shift in understanding awakens in us the vast potential of interconnected ideas and new habitats in an infinite universe operating under a higher ethic than that confined by human beings. The universe would perhaps face the same dilemma should it become unreliable, unperceptive, and finite, ultimately not with the potential to expand and evolve.

Now with infinite potential under, $\infty\Delta C(\text{morality})=\sum(\infty\Delta a\approx\infty\Delta b)$, life as an off-chance event and extinction theories become less threatening in the perpetuation of an interconnected life. Such life is driven to be principled both internally and externally.

If any universal morality is not based on tolerance, selflessness, respect, truthfulness, and with its main interest an interconnected progressive cognitive wellbeing, what else can it be aimed at, from where we stand as a perceptive *a* in full regard of an interactive *b*? I found no reasonable answer to this, except to question the current human condition and rebel even more fervently against an antiquated restrictive version of evolution—only capable of inflicting ongoing needless suffering with its despotic mindset.

* * *

I postulate here that any responsible search for morality in our actions is made less equivocal and more pragmatic if seen as a highly interconnected, dynamic, tolerant, and pliable *universal idea-making process*—operating in a perspicuous Physical sphere of reasoning. A morally conducted pliable perception is essential in our search for a 'better world,' as probed by the late Sir Karl Popper, and is now further supported as a *vital* drive. Our objective presence is transient; however, our ideas (also ephemeral) can serve, subject to its morality, to change our world and our perception based on reliable concepts with its place in the universe for future generations to expand on.

A meaningful better world simply cannot be achieved by a delusional society masquerading behind a misinterpreted version of Darwinism or alternatively trying to find escape in false belief systems. We can perhaps also begin to sense how striving for a universal morality with fair and workable social systems is not idealistic but inevitable as part of our evolutionary design and destiny. We have no choice other than following and advancing a progressive universal ethic, *together* and sooner rather than later, unless we are willing to surrender to a non-receptive barbaric unworkable evolution. This may perhaps be why higher 'alien' intelligence, recently becoming an entirely plausible theory among some astrobiologists, remains alienated.

I hope it can be clearly seen here that my argument is philosophically neither in support nor in denial of an embodied cognition or a narrowly defined acceptance of traditional cognition. I claim that human cognition (and morality) is simultaneously a dynamic, skull-bound, functional evolutionary status and vitally and intricately interconnected to body and habitat. It is furthermore misdirected by false beliefs and political dogma or any other means of external control and false beliefs. Both cognition and morality demand plasticity, genetic adjustments, interconnection, and existence failure in a dynamic interactive environment, with moral conduct a vital and essential ingredient for pragmatic outcomes.

I further argue here for the urgent need to distance ourselves from a generalized reductionist view of naturalism centered around survivalist concepts and reproduction, in exchange for an evolutionary process seen as an active interconnected, highly moral, falsifiable perceptive process. Moral behavior now becomes obligatory rather than facultative. Reproductive success and status is now seen not as a callous solipsistic genetic triumph but merely as a method used for idea propagation, with all life as valued key ingredients. An interconnected perception, in turn, not only drives evolutionary cognition now but constantly adjusts itself and the genetic tools it is derived from. We in turn have a progressive, adaptable, purpose-driven evolution with infinite potential to overshadow extinction. From this evolves a much more dynamic and pragmatic ontology when compared with the previous narrow confines of an ontology reliant on a cognition inebriated by an unperceptive disregarding natural selection based on the improbity of 'adapt, survive, or die' principles only. We also have more scope to develop a concordant society with ameliorated morals and progressive knowledge.

Important concepts in developing such a universal ethic include:

- Pliancy
- Interconnection
- Respect for failure
- Evolutionary cognition as a rhizomatic 'universal' increase in complexity based on establishing *reliable* links.

Awareness and understanding of these concepts can also extend into a motive and model to avoid the manipulative and restrictive impact of biases in business and healthcare, and many other fields of research and application. This can fuel a more progressive Popperian formula:

P1→TT→EE→P2

Here, Tentative Theories (TT) and Error Elimination (EE) emerge from one problem (P1) and will flow into the continuous creation of another problem (P2, P3…and so on). We now have a progressive perspicuous evolution set in infinite spheres of perception and ideas based on truth and trust. Tentative theories exist in our Physical sphere of reasoning, always evolving new problems for error elimination in the Logical sphere of reasoning. All these evolving problems and tentative theories interconnect in the different spheres of perception. Similar to networking constantly adjusting neurons by means of dendritic connections in a rhizomatic pattern, harmonizing the interconnection of a unison of neurons, so also do our spheres of reasoning evolve plasticity and pragmatic complexity in the Physical sphere of reasoning.

We can now understand and begin to progress a morality embedded in an embodied interconnected organic evolutionary cognition, both intrinsically and extrinsically. Such a morality accommodates a dynamic evolutionism, beyond the narrow confines of body and mind, yet simultaneously totally interconnected and committed to its perceptive internalism.

## 2.3 Physical Sphere and Logical Sphere of Reasoning and Metaphysical Interactions

We have mentioned the pliancy of cognition in our adaptable idea-making process. We can perhaps sense the inevitable need for some form of security in our cognitive process, albeit ephemeral in a more universal context of things. Such temporary 'certainties' are vital to confront the pressing realities of daily life as organisms try to survive challenging and constantly evolving habitats. Simultaneously, an organism must remain focused on workable ideas within an interdependent cognition. Because of these interconnections there are shared interests in pragmatic outcomes on all levels. Inevitably, with a rapidly growing technocratic society where an 'Internet of things,' blockchain technology, and numerous ideas set root to a flood of Internet traffic, social media, and data (having already exceeded one zettabyte in 2016), we need to employ enormous effort to sift through rational ideas, pseudoscience, and gibberish. We should also avoid hiding behind natural science and becoming too reductionist or dogmatic, even in our views here. Untenable ideas are further made complicated if we consider brain function (cognition, C) as an adaptive and constant change within a change with individual variability, meeting the criteria of $\infty\Delta C=(\infty\Delta a\approx\infty\Delta b)$—where pliant workable ideas constantly change based on interactions on all levels between the observer *a* and the observed *b*. It is important to note that *b* includes all aspects of the environment, including new objects (some manmade) and ideas that *a* is exposed to. With an escalating and constant interchange of knowledge and new *products* released almost daily, principally targeting a more confused market, it can be difficult for consumers to always make informed decisions. We now have more *a*'s connected to more rapidly changing *b*'s—vitally dependent on a simultaneously progressive morality functioning as a network to make informed choices. The moral demands and complexity also grow as the network expands.

* * *

Survivalist ideologies driving most economies and corporations today, with the principal motive of how to retain profits in a technocratic society, also do not help much more than to threaten the science and morality of a society reliant on the candid pragmatic plasticity of its open knowledge. In view of my argument so far, set in a cognitive flow within an equally malleable environment we can see, not only is the enormity of the task for the supporters of trickle-down economies and the reductionism of 'what pays' perhaps unrealistic, but also in urgent need of transformation. The potential utilitarian value of object permanence, AI, and robotics is not in the scope of this text. While the potential of reductionism and the moral implications with simultaneous escalating unemployment figures is concerning, the utilitarian value of robotics in our advance simultaneously cannot be ignored. However, besides the inability to match the plasticity and uniqueness of the organic brain to adjust more liberally to escalating change, we should also consider the potential loss of an infinity of unknowns and new potential that the unfolding universe continuously creates. It evolves our knowledge as an interconnected concern—while evolving and interconnecting reliable information. We can see the two interconnected levels of evolution here. We should also accept the current idea that genetic diversity in the animal and plant kingdom has been established by means of natural selection. This process can be dissected down to the physical realities of RNA, DNA, nucleotides and genomes, and various chemical elements to intercommunicate, with reproduction and survival as obvious needs. This idea secured itself firmly in our expanding ontology like fire, the wheel or numbers as the building blocks of mathematics in our *Physical sphere of reasoning*.

It has likewise become a completely temporal issue to accept the dynamic pliancy and versatile responses of protein receptor-sites and exchanges involved on a cellular and physiological level. These exchanges create interaction between cells, organelles, and organs, all performing various physiological functions receptive to progressive habitat and surroundings. Such objective realities are continuously studied, well documented elsewhere, and constantly improving our knowledge about what we are and how we function and may potentially improve our lives in search of a better world. It has obtained scientific credibility beyond dispute and has firmly entered our Physical sphere of reasoning (PSR) as an essential part of a progressive empirically based epistemology. What remains open to discourse is why and if there is an *ultimate* morality in this struggle to survive, perceive and diversify, or even some 'higher' transmogrification in it all? Perhaps this must remain an ongoing open query, coexisting with the need for infinite change and unknowns as essential stimulants to evolve while co-evolving our morality.

Based on my argument here, I shall seat the *why* we struggle to survive (traditionally metaphysical) in our Logical sphere of reasoning, as propagating and nurturing interconnected progressive ideas, set in a sound ethic to progress our perception, understanding, and morality. We cannot evolve without escalating our perception. Simultaneously, to advance rhizomatic interconnections with increasing complexity we need trustworthy interconnections in a sound cognition, interlinked to a sound epistemology.

In awe of this enormous task and aware of the active malleability of our cognition in its equally vulnerable and changing habitat, we can see how any attempts at reductionism, fixed reality, or object permanence, as a base for a normative, may be unrealistic—while we constantly evolve morally. We are rationally seen as no more than a changing idea within an evolving idea. We also clearly, however, cannot survive without the security of a principled perceptive sphere where workable ideas are harbored to guide us. In this Physical sphere of our existence with the primary value of improving our quality of life and chances to survive, we also serve to propagate new ideas to benefit future generations. We therefore need a dependable Logical sphere of reasoning to present candid ideas to our pragmatic Physical sphere of reasoning. I am duty bound to next define these spheres of perception (Physical and Logical) better, while steering well clear of being accused of reductionism.

* * *

With cause and effect being interchangeable and exposed to continuous idea-making, subject to testability and falsification, these definitions should also not be culpable of restricting causality. We also need to have a clear approach to the metaphysical in the context of this text, not necessarily in a traditional philosophical context.

Subsequently, we define the *Physical sphere of reasoning* as an interactive perceptive sphere, or mental sphere, where perceived realities (workable ideas) enter our thoughts and remain established due to their value in improving our quality of life and ability to survive. Such workable ideas as recognized by our cognitive abilities will often be supported by evidence in mathematics, physical sciences, and nature as measured relative to other objects. We shall use a few examples below, variable depending on *what* you are, *who* you are, *when* you are where you are, and most importantly constantly interconnected to the security of a *pliant* Physical sphere of reasoning, periodically shifting ideas. These spheres subsequently reliably operate and interact on all levels of perception regardless of species.

Diversities set in a realistic Physical sphere of reasoning offer pragmatic security and we can use any example that comes to mind (note the value of numbers):

The angles of a straight-lined triangle add up to 180 degrees in algebra; $HO + CO_2 \rightleftharpoons HCO_3 + HO$ in a given bio-environment (biochemistry); $E=mc^2$ (physics). Practical examples not in need of much further evidence or argument include the following: A fall from a 25-meter-high building operating under Earth's gravitational forces onto solid concrete will inevitably cause bodily harm as a 70-kilogram human, most likely not as 3-milligram ant. Eating 75 grams of chocolate if you happen to be a 2-kilogram chihuahua may be lethal due to its theobromine content, but if you are a human most commonly a pleasant experience. If you are a lame zebra on the plains of Africa, a hungry lion will see you as easy prey, but not in London Zoo where a vet will attend to any malaise and so on. In brief, clearly all physical realities if not pragmatically perceived and related to by an organism will lead to its demise, harm, or a predictable (to some extent) experience, depending on the organism and variable not only to *what you are*, but also *where* and *when* you are what you are.

To a slime mold most of these concepts belong to the Metaphysical (or Kantian noumenal—see later) on a perceptive level, but as a human are solidly placed in our Physical sphere of reasoning.

* * *

The sphere of *Logical reasoning* is a boiling-pot of ideas and thoughts about objects or non-objects. Here, ideas are received and evaluated with some uncertainty and perceived by an evolutionarily adjusted perceptive-mechanism (a pliable DNA/RNA), with much variation in cognitive means between species, and even individuals of the same species. In the Logical sphere of reasoning objects and non-objects or ideas about them are, after evaluation, merged into the Physical sphere if realistic and backed by evidence. The Logical sphere is a lucrative sphere where abstract concepts and also creativity may originate. Concepts such as Darth Vader and fairies and, less notional, our speculations about alien life, all reside here. Embarking on a project in cancer research will be less abstract on initiation, with some guidance from previous research in our Physical sphere of reasoning. The outcome, however, remains in our Logical sphere of reasoning until the research presents us with sufficient workable evidence to be presented to our Physical sphere of reasoning.

The Logical sphere is also the fragile domain where ideas are dependent on a *healthy* cognition and sound logic and may be easily manipulated and unhinged. For example, falsification of research results for personal gain and manipulation of ideas by media or unproven beliefs, or more evidently a dysfunctional cognition—perhaps daily heading off to work with my light-saber, convinced I am Han Solo. Sadly, many so called 'remedies' in healthcare today hover in this uncertainty of the Logical sphere instead of the Physical sphere and yet are promoted by those hungry to market novelties. Numerous belief systems and religious practices are also Logical sphere activities with no concrete evidence of the core belief they are centered around. We also have to query the social value and sense in defending one untenable belief as more correct than another, or one religion more authentic than another. Likewise, vaguely referring to everything as designed by an abstract Creator in our Logical sphere can precariously be acceptable only in metaphysical terms, with the existence of this manmade core belief having never been proved.

In a society with growing mental health issues, a few other examples of a badly orchestrated Logical sphere would be to move away from a food source when close to starvation, substance abuse, killing another person because your religion urges you to do so, eating a plant without knowing whether it is toxic, failing to move away from the potential bite of a snake, politicians creating unsubstantiated fears based on scanty evidence to gain support; or ignoring the prospect and potential impact of climate change (when it is a real threat), blindly following an outdated economic system, or consuming feckless supplements or medicaments with uncertain 'health benefits.' The list is endless, and we have to question why the brain is subject to adhere to such illogical conclusions and false beliefs with at times large-scale negative impact, clearly unacceptable to a sober Physical sphere (more on this when we discuss mentalizing).

* * *

In logic, when told A is bigger than B, and B is bigger than C, we can conclude that A is bigger than C in our Physical sphere of reasoning without having to visualize any of the objects. Using simple deductive inference, initially we function in our Logical sphere of reasoning, and place an arrived-at conclusion in our Physical sphere of reasoning. With 'rational pliancy' in the Physical sphere it may become established as tacit knowledge (without much thinking) for future use. We should note, however, that if A was a large block of ice and B a rapidly growing seedling and C a chicken egg about to hatch, upon time-interrupted returns to these three objects all our Physical sphere perceptions would have been erroneous and would change. We can see how limited things would be in our Physical sphere without exposure to a dependable *evolving epistemology*. This reemphasizes the importance of evolving trustworthy knowledge, and in today's data world, 'sober data' as a platform for a pragmatic knowledge to continue evolving.

We also need to remind ourselves that cognition is a fluctuating temporal activity perpetually exchanging ideas between a pliable Physical sphere of reasoning (PSR) and an even more malleable Logical sphere of reasoning (LSR) with both spheres confronting vast metaphysical (M) uncertainties, and also new potential.

We generally would not give much thought or employ complex measuring techniques to confirm that a tennis ball is smaller than a soccer ball. Similarly, it has become tacit knowledge today that the moon is smaller than the earth and the sun larger than the earth using mathematical calculations as well as landing on the moon to support this belief. Without mathematical calculations and a Copernican universe, we would most likely not have landed on the moon or be able to conceive the vastness of the universe today. Likewise, the confident security that mathematics offers our Physical sphere of reasoning is based on the fundamental evolution of numbers and our interpretation of object, where object presence=1, 2...and so on, or absence=0. Furthermore, prior to developing the 'language' of mathematics and the ability to count and calculate distances and values relative to numbers, we may be forgiven for having entertained the possibility of the Earth being flat. It should already be clear that the movement of ideas between the Logical sphere and Physical sphere of reasoning depends on what you are, who and where you are, and importantly *when* you are there and the state of an evolving epistemology. One billion years from now our sun, if predictions are right, will turn into a White Dwarf and would not have the same format as today and neither would our solar system, or us undoubtedly have the same cognitive status and perception or presence, should we still exist or perhaps dwell elsewhere in the universe continuing to evolve new ideas. Furthermore, numerous undisclosed concepts now still in our Metaphysical sphere would have entered and changed our Physical sphere and would have come and gone in our Logical sphere, and the Metaphysical also evolved accordingly. The perception of the sun and the moon to a human, chimpanzee, humanoids 5000 years ago, or an earthworm today are all vastly different. Even from an epigenetic perspective it has had changing effects: Ice Age and now impending global warming (we suspect).

A striking resemblance between this spherical interchange of reasoning and cognitive evolution is now noticeable. Pliable genes (and alleles) express genetic ideas in a genetic Logical sphere and can either be expressed as functional ideas in a genetic Physical sphere and then may act as *pliable* blueprints for expansion, or fade and return to a genetic Logical sphere or metaphysical (extinction). Some allele combinations have become firmly established in a genetic Physical sphere and essential for the organism to confront and adapt to the physical realities of its changing habitat because they meet ongoing pragmatic needs. Others may hover in uncertainty for long periods, with seemingly no utilitarian value, the equivalent of a genetic Logical sphere. Some slip through with both benefits and negatives, such as the historic and often-mentioned example suggested by biologist Julian Huxley in 1964. With much newer research in recent years, thanks to giant leaps in genetics, the historic example given by Huxley (*Nature* 204(4955), pp20–1 [October 1964]) claimed that schizophrenics are "also concurrently considerably less sensitive than 'normal' persons to histamine and as a result less prone to suffer from operative and wound shock, and not suffer nearly as much from various allergies." We can see how the trait for schizophrenia, unless favorably promoted and selected because of some other need, will to some extent be drowned in the complexity of various outcomes and genetic combinations. Should it somehow become established, let's say due to a mass increase in serious allergies, the unfavorable social outcome will become more apparent. It is, however, still unlikely that reliable links can be established in a perceptive network based on this trait (albeit allergy free) to grow pragmatic knowledge, and more likely that such a trait will eventually be even more rapidly diluted to return to the Logical sphere or metaphysical due to its unreliability in establishing trustworthy connections. So, we see that the Physical sphere (both in cognition and genetics) evolves from reliable interconnected and vested long-term perceptive interests, rather than narrowly set individualized material aims and deceptions.

We now have an evolution with genes and epigenomes (chemical substances surrounding the DNA helix), progress and adaptability based on a complex interconnected perception, and progressive morality (principled links) in an interlinked evolving perceptive network. We can confidently place this in our Physical sphere of reasoning.

* * *

What remains as a concern in society today is how the two main controlling concepts, religion and our current economic system, are both functioning mainly in the Metaphysical or Logical spheres. With significant impact and creating conflict in the Physical sphere, both are yet centered around equivocal and unproven core-beliefs. In the case of religion, the various interpretations of a deity and in the case of the current economic system, the belief that uncertainty in markets, with strategic maneuvering of resources, are our best and only options. Both systems are imperceptive and reductionist and guilty of promoting segregation and isolationism. It is relatively easy to see when a society sets its core values on beliefs with no clear evidence and much uncertainty, that such a society will be more prone to strife and a faltering morality, besides being stigmatized by its ignorance. Inarguably, few can doubt the clear evidence that a hot plate will burn your finger or that running in front of a speeding truck is a poor decision, if one values life overall. We can be forgiven for expressing our concern here that much of the mental disease, deprivation, corruption, and cruelty of our current world can perhaps originate from disillusionment when we sense that the core values we base our main structures on may be false or manipulated and may not meet the requirements of a progressive society based on a sober and principled Physical sphere.

* * *

Returning to healthcare, where we are confronted by a bombardment of new (at times conflicting) ideas in medicine today, we are obliged to momentarily turn to a technique that may superficially appear very attractive. As a method to create verisimilitude in our physical sphere, Evidence Based Research (EBR) currently serves as the gold standard in healthcare and acts as the principal umbrella used to screen the safety and efficacy of techniques and all the medicines we consume. Be reminded, during our analysis of EBR, both the genome, epigenome, and the idea-making process on all species and sub-species levels are persistently striving for a harmonious $\infty\Delta a \approx \infty\Delta b$ *within* an infinite exchange. The pliability of the epigenome and its diversity is also now well recognized. We simultaneously have a constant flow of exchanges between Logical and Physical spheres with new medicines and techniques constantly replacing existing ones. This delicate need to replace ideas with new ones is a process dependent not only on workable and trustworthy ideas but also on honest and scrupulous perception. Currently what also hovers over the whole system is the prospect of profits to be made.

## 2.4 The Limitations of Evidence Based Research and Medicine

Evidence-based medicine is defined as "the *conscientious, explicit, and judicious use of current best evidence in making decisions about the care of individual patients*" (D.L. Sacket et al., *BMJ* 312 [13 Jan 1996]). Left unchanged since this early definition, it should at its maximum be taken as a valuable attempt to help create acuity in our Physical sphere.

Dissecting the above definition, we can sense semantic complications at first glance when we define *explicit* as implying clear and complete, unequivocal, and having as its antonym *vague*. *Judicious* implies wise and thoughtful or intelligent and having as an antonym *stupid* or *ignorant*, so we immediately face major relativities and impediments here. As an example, it is entirely possible for mistaken use of current objective values to come up with scientific breakthroughs—Pasteur comes foremost to mind with his discovery of penicillin, also the accidental discovery of the amino acid Lysine's effect on Herpes viruses, and many more. Likewise, it is entirely possible to judiciously apply healthcare based on currently accepted EBR recommendations, proving one or two years later to be based on erroneous concepts, even causing harm. It may be a shock to know how often this is the case; and now pharmogenomics as an emerging field is raising even more concern regarding this rather delicate issue. Furthermore, judging what is wise, thoughtful, or intelligent has become exceedingly complex due to biases kindled by personal gain, status, culture or religiousness, resulting in mass manipulation of the Logical sphere of consumers. And this we certainly should not tolerate in an already delicate arrangement as explained so far, especially when it comes to serving as the core guide in caring for our health and wellness when we are at our most vulnerable.

With all effort made to be ethical and honest, healthcare workers under a sworn oath are operating under the dogma of EBR while treating a diversity of life and disease, but we can see how they may be innocently vulnerable to be accused of judiciously applying *stupid* ideas (using an antonym for *explicit*) at times. Any clinician today can call to memory a number of costly and heavily promoted drugs or techniques used a few years ago, previously strongly supported by EBR, now discarded or in disrepute due to either lack of showing any significant *in vivo* benefit, or at times developing unforeseen side effects or even causing death. These concerns escalate where scanty past EBR supported the ongoing use of drugs with little benefit to the patient, or certain 'natural' remedies and nutraceuticals, and all goes haywire from there, lured by profits to be made in an open market swaying feckless consumption. This potential to promote a pseudoscience, with complete disregard of the environmental impact and financial burden to an already overstretched consumer, can be driven by profits to such an extent that even some rather minuscule potential benefits are inflated as having major health benefits. Clinicians, trained in the science of life with healing as their main motive and with an understanding of their patients' vulnerability and disparate emotional needs, are pledged to focus less on feckless medicaments and more on pragmatic remedies, but how do they do this, set in all this complexity in our current consumerist society?

We should further scrutinize the process of EBR functioning under the influence of certain sociopolitical structures very carefully before we place it too close to the core in our Physical sphere (see Figure 1). Often based on a pyramid schematically, the path begins with biomedical research and then goes as follows:

The shape of the pyramid symbolizes the process of testing innovations in healthcare and eliminating those that lack merit. Financial stakeholders and government or corporate employed scientists mostly have the deciding vote here with immediate biases already setting in, selecting those medicines most likely to quickly deliver large short-term financial returns or match the investors' interests. The broadest base of the pyramid represents the initial testing of innovations, which usually occur in laboratories that are today mostly privately funded or owned by large financial concerns with their own team-leaders as the experts leading the study, this further impacted by political influence and beliefs. This foundation of the pyramid is subsequently heavily influenced by personal world-views, profits above all and politics. This level is vulnerable to accept any anecdotal evidence that may support such an investment.

In turn this is also where some products or ideas are outright rejected due to financial viability with the main decision-makers not necessarily trained in the science concerned with the testing or motivated by and familiar with the suffering of the victims of specific ailments—potential benefits may then be lost in the metaphysical here. With such a shaky base, some products or processes with 'merit' (with profitability a strong factor) then undergo field trials; these initial studies aim to assess toxicity and to estimate efficacy based on laboratory and subsequently animal studies. Recently even the animal model is under question, as scientists realize more and more the subtle cellular and genetic-level variability under differing conditions, even in the same species. Some may benefit, but a small number may potentially even die when taking the drug. We need to recall our concepts, discussed in earlier chapters, of mobile DNA, constant change and variables and who, what and when, and now also pharmogenomics.

Principal stakeholder bias is inevitably directed at proving efficacy once money is invested. Many innovations do fail, but a few merit more definitive testing in large controlled trials with important clinical endpoints set; some minor side effects are often overlooked here because of vested interests. It is only when judgment is applied to such studies (another area of bias in interpretation here) that they are successful and serious efforts at dissemination and application are forwarded. Already bias (even fraud) may set in here as well. Increasingly, behavioral interventions, surgical procedures, and alternative approaches to the organization and delivery of care are being subjected to similar evaluation with similar weaknesses and biases.

The pyramid is then classically based on **levels of evidence**, as follows:

- Category I: Evidence from at least one 'properly randomized' controlled trial (mostly corporate funded)
- Category II-1: Evidence from well-designed controlled trials without randomization (biases may exist in electing facilities and workers who may benefit from these trials)
- Category II-2: Evidence from well-designed cohort (sharing the same characteristics) or case-control analytic studies, preferably (note the wording) from more than one center or research group
- Category II-3: Evidence from multiple times series, with or without intervention or dramatic results in uncontrolled experiments (subjectivity strongly enters here)
- Category III: Opinions of respected authorities, based on clinical experience, descriptive studies and case reports, or reports of expert committees. As mentioned in our opening argument two thirds of these authorities may have personal interests in the outcomes, generally financial or as affecting their own research and publications.

Consider the interests in such research and outcome based on the parties attracted to such an emolument (also keep in mind that the final approval for release is still vetted by politicians and governmental bodies). We can list them as:

- Principal financial stakeholders and pharmaceutical companies
- Researchers gaining reward and academic recognition
- Academic culturism and pressure to secure ongoing funding for future research
- Patients and the family of patients—minor input unless as recruits in a study group

- Health insurance companies
- Politicians and the business fraternity
- Support or concerns backed by religious or other belief systems
- Current ethics committees, inevitably and without exception also with some leaning toward one of the above based on background, support, education and belief.

Although the principal motive is germane, it can be seen how EBR may still be vulnerable to many weaknesses.

Now further pressurized by new evidence of the *transgenerational genetic impact* of many drugs and chemicals, concerns are raised to another level. For the manufacturers and distributors of drugs, especially where backed by scanty evidence and unclear value, the genomic effects that can last for some generations may be a significant concern.

We may perhaps now ask:

1. Is it at all possible for a universally beneficial and honest result to ever emerge, considering the complexity and diversity of this conglomeration? Perhaps this is our only and best option?
2. What is the relative impact of the different parties on the project? Indisputably, under current systems funding has overwhelming control.
3. Is enough credibility given to quality of life in such studies?
4. Is the environmental impact considered at all when manufacturing and marketing products with scanty evidence or minimal impact (as long as they seem to cause no obvious short-term harm to the consumer)?
5. Do these results justify research spending on an obvious larger-scale long-term genomic and environmental upset? The long-term genomic impact is only a recent new concern, as mentioned.
6. Do false belief systems, politics, culture and nepotism still have any impact?

7. Do the threats of poverty and unemployment under corporate-focused wealth in recessionary times make it easier to buy or swing scientific evidence?
8. Is there an ethic involved here at all and, if so, what is its main directive (we may be forgiven for suspecting it is profiteering)?
9. To what extent does political power (Foucauldian impact) and corporate muscle affect results? Is a reductionist Darwinian approach still at play here?
10. Is medicine a science after all, or do we redefine medicine or science and in view of our current knowledge relocate it to the LSR?

Profit motivated and strengthened by studies to cover up the major side-effects and with focus on narrowly defined benefits, inflated merits now become the main target. There certainly are many more concerns that can be formulated around EBR but, this aside, the fact is that the consumer today is bombarded with an array of expensive drugs, natural remedies, cosmetics, and nutraceuticals, all with the promise of improving our health and yet with growing confusion and concern that many of them may *not* have any significant overall benefit, in either short or long-term healthcare outcomes.

* * *

Another major and much ignored perturbance is the still complete disregard of the potential environmental impact and future transgenerational genomic impact in using these agents. The *principal motive now has slowly shifted from pragmatic knowledge to heal, to how to accumulate profits.*

Clearly this is wrong, and it appears obvious that EBR may be misdirected when afflicted by profit-seeking behavior and personal interests, rather than more goal-directed cures with long-term gains in a vested genomic outcome. Some motivations may be as egocentric and short-sighted as status and quick financial returns. The idea of trans-generational genetic and environmental impact is also not currently included as a factor in any pre- or post-research market surveys. Neither is pharmogenomics gaining appeal when investors stand a chance of losing some of their current market, should this avenue open up. When genomics can be employed to check for individualized drug efficacy or side effects prior to use of drugs, it is predicted by some scientists that somewhere around 30 percent of individuals will have no benefit and some will be exposed to serious risks in taking some of these drugs. The latter figure is based on recent work by James Kennedy, at the University of Toronto in Canada, who heads the Tanenbaum Centre of Pharmacogenetics at the Centre for Addiction and Mental Health. Pharmogenomics has become quite workable and relatively inexpensive (compared to some patented drugs), now that the human genome and the genomes of some of our more popular pet animals have been fully mapped. Evidently then we need to change the current ethos in healthcare, starting from the base of the pyramid in EBR. If we can manage to reason in an *un*manipulated interchange of concepts between the Physical, Logical, and Metaphysical spheres of our reasoning, we can perhaps hasten our evolutionary destiny and heal more pensively, as we should and are designed to do with a more reliable EBR then operating in clear PSR.

## 2.5 The Imperative of an Interconnected and Progressive Physical Sphere of Reasoning (PSR)

There is clearly an urgent need to filter reliable, pragmatic, and truthful ideas through to our Physical sphere of reasoning to support a verifiable model of evolution and epistemology.

Evidently, we may have to assimilate numerous ideas in our Logical sphere before we can accept them in our Physical sphere of reasoning. Turning to physics in an attempt to restore a bit of confidence, we can consider a quark as a simple example. Quarks are currently known, together with leptons, as the smallest subunits of matter existing in the universe. According to the scientists at CERN, they are close to discovering many more subunits. We can then, based on current knowledge in quantum physics, at least entertain quarks in our Physical sphere of reasoning and the potential of discovering new subunits in our Logical sphere. It is possible for us to consider prospective subunits in our Logical sphere with the likelihood that new subunits may be uncovered. We can place this closer to the core of the Logical sphere of reasoning (see Figure 1). To prove the presence of weaker or stronger presence in the Physical sphere or Logical sphere of reason, we can use any other example we want and pencil it in closer or further away from the core of the sphere.

From a healthcare perspective again, many nutraceuticals, traditional, and natural remedies currently move between the spheres but essentially belong to the Logical sphere of reasoning. Some natural remedies belong to the outer realms of the Logical sphere, even bordering on the metaphysical, but are still commonly consumed. Excessive consumption of certain herbs and plants without enough evidence but supported by either a traditional belief or the discovery of a potential beneficial physiological action of an element found in such natural remedies, is now commonplace. A classical well-known example here is the recent support of the role of antioxidants in cancer. These were first published as preventing cancer, then—perhaps less well promulgated—as having a pernicious effect in helping it to spread. Overstated claims resulted for a period, with blueberries even ending up in pet food as a cancer-preventing diet, and much overinflating of health benefits based on such disjunct and rushed deductions. The more recent metastatic potential of antioxidants was backed by findings in work done by James Watson (of DNA-helix fame and Nobel laureate) and his team. Likewise, with the potential benefits in taking glucosamine (an amino sugar) for arthritis sufferers, again we find conflicting support in some studies, still, however, resulting in a booming market around this still debatable evidence. With no mention of the ecological disruption, as this substance is sourced from green or red-lipped mussels or shark cartilage, or the future genomic impact of these mass-distributed substances in both human and veterinary healthcare, we operate perhaps dangerously here in the Logical sphere of reasoning.

Commonly used pharmaceuticals today, with no sufficient in-practice genomic screening in place and no clarity around their long-term genomic impact, consist of an extensive list; statins and non-steroidal anti-inflammatory drugs stand out, based on familiarity and high usage. Besides this lack in sound knowledge about their long-term genomic and environmental impact, their minor benefits when compared to those gained from only improving dietary and lifestyle choices on the epigenome are vastly overlooked. Casually noted by some research in dermatology recently, statins were possibly linked to an increase in basal cell carcinomas (a skin cancer) in women (*British Journal of Cancer* 114(3), pp314–20 [2 Feb 2016]). These figures are difficult to interpret and open to further investigation but still noteworthy and hard to ignore if perhaps you suffer/ed from basal cell carcinoma.

Emphasizing the need to focus on physical activity, stress reduction, creation of more stimulating and healthy environments, and, of course, a healthy diet (low in meats or vegetarian), is not very lucrative to promote or include as control groups in studies mostly funded by pharma. Such simple and sensible methods are, however, free of side effects and kinder to the environment and can have an enormous beneficial impact on the epigenome and our health. They will also significantly reduce the distribution of chemicals targeting production-animal and human medicine markets.

* * *

Turning to a completely different aspect, geophysical events may potentially cause extinction and significant shifts between the Physical sphere, Logical sphere, and Metaphysical. An example is the extinction of dinosaurs. Here a singular event, proposed as a meteorite strike, eradicated an entire sub-group of animals from a Physical sphere to re-emerge in our current Logical sphere, 60,000 years later. In the futuristic scenario where life as we know it on Earth is eliminated by, perhaps, a supernova explosion this time, our current Physical sphere would similarly disappear and remain in fossils and recordings of how we shaped the world around us, and perhaps DNA or digital recordings discoverable by future perceptive beings. Exactly how we lived will remain a Logical sphere activity to entertain future civilizations. Regardless of where we stand in time, sphere or belief systems, we are all temporary Physical sphere participants interconnected to an expansive perceptive network, before we all return to the Logical and Metaphysical. Only our ideas (thoughts) are continuous with the infinite sagacity generated by this evolving interconnected matter.

For explanatory purposes, we may continue with dinosaurs. Albeit well removed from healthcare, this example is well suited to explain the impact of *time*, *what*, and *how* on our Physical sphere. Dinosaurs have strong objective evidence in carbon dating and piecing together their fossils, so they certainly are more prominently placed in our Logical sphere than concepts such as Darth Vader or fairies. Note, however, that neither fairies nor dinosaurs exist in our daily lives and yet they both entertain our Logical sphere to some extent but not our Physical sphere, unless science one day perhaps presents us with a Jurassic Park. We also must realize that *how* they lived is based on fossil recordings and carbon dating and is merely a 'how we think it was' scenario, based on piecing together current evidence. This evidence does exist in our Physical sphere, but the reality of a living dinosaur does not. It is important to realize that *time changes everything*, even our thoughts on this.

The pragmatic value of dinosaurs and extinct species in our Physical sphere today is limited perhaps to better insight and understanding of evolutionary roots and extinction theories, or perhaps for its entertainment value, as with dinosaurs and fairies. The more imperceptible idea of fairies, however, lacks any objective evidence to prove their existence and probably never will gain support (unless perhaps humans are genetically modified, or aliens arrive presenting themselves as small humanoid creatures with wings). Still, they do occupy our Logical sphere of reasoning in fiction and stories we tell our children (how can we ignore Tinkerbell?).

* * *

We should use one more historical example in biology for our purposes here to introduce the concept of **spherical interchange of ideas** as a pliable yet rational aid to arrive at logical conclusions in a reliable Physical sphere. Perhaps the more moralistically science-related intricacy of the historic **preformationism** can be used.

If we lived in the time of one of the first great embryologists, William Harvey (1578–1657), and his then contribution to our evolving epistemology, evidence at the time would have presented us with **ovism** as a plausible explanation of how new life forms. Harvey well supported the idea (and social needs) that all organisms start life with all the parts already formed and then simply grow bigger and bigger from there. Before we judge this (ovism) perhaps as an absurd concept, we should consider why this was so.

Marcello Malpighi, another prominent contemporary scientist, with an invaluable contribution to our progressive knowledge, was the first to turn the then novel and invaluable discovery of a microscope enthusiastically to biology and so opened up a whole new world and universe for scientists to discover. Prior to this the microscope was used primarily for star gazing by its inventor, Antonie van Leeuwenhoek (1632–1723). Ovism then would have been set in the Physical sphere of reasoning as a solid guideline in a developing natural science of that period. Van Leeuwenhoek in the same era soon also, armed with his new invention, started looking at the microbial world and replaced ovism with the equally unrealistic theory of **animaculism**. Here the embryo was seen as preformed and carried inside the spermatozoa. This misogynistic misconception had a solid spot in the Physical sphere of scientists at the time with an impact on morality that lasts until today. All this fully supported the doctrine of *emboitement*, influencing the Church, charged as the director of morality, to condemn abortion at any stage as the killing of a fully formed little human being. Today none of the above can sensibly be placed in any Physical sphere of reasoning or used in defense of an anti-abortionist stance anymore. A more realistic and defendable approach for anti-abortionists, in an era of stem-cell research, would perhaps be to consider respect for the interconnection and sacrosanctity of all life, from the unicellular level up. With a concordant origin, we do now have sufficient evidence of this responsibility.

From the above example, we can clearly gather how insecure both our scientific foundations and belief systems are, and the absolute need for pliancy and responsibility in a Physical sphere to carefully nurture a progressive pliant universal perception that harmoniously adapts to change and time.

* * *

To recollect our postulate so far:

- The Physical sphere, Logical sphere, and Metaphysical sphere will depend on who and what you are, and *when* you are what you are. It is individual, species, and *time* linked and constantly *changing*.
- The same Physical sphere can also influence the Logical sphere differently in different individuals of the same species (intra- and inter-species, non-specific) in the same period.
- New discoveries can cause ideas in the Logical sphere to replace an existing idea in the Physical sphere—returning earlier theories to the Logical sphere or eventually the Metaphysical.
- Inter-species effects—the effect of one species on another—can also be marked. As an example, some species have gone extinct due to human activity.
- There are some Physical sphere certitudes that apply to all known lifeforms on earth regardless of species. As an example, it is not possible for any creature to live in temperatures above 150 degrees Celsius, as far as we know. Even for thermoduric microorganisms, with the highest recorded thermoduric living

> microbe currently known, found at around 121 degrees Celsius, such temperatures would spell their demise. We can perhaps with much uncertainty entertain in our Logical sphere the thought of some lifeforms elsewhere in the universe evolved to exist under higher temperatures.

We can see the complexity and variability of security, even in the Physical sphere, and yet we can start sensing the possibility of creating a Physical sphere of reasoning where some ideas can confidently proximate with the forever elusive core truth—while co-evolving with the LSR and MSR.

* * *

One more example perhaps before we continue on our journey. To fly to London today trying to land an airliner, simply by going on computed mathematical calculations around continental drift and where we estimate London should be today after a hundred million years of continent drift, would obviously be ludicrous with 300 people on board and their lives at stake. An expansive pliable Physical sphere must combine *current* pragmatic ideas with *time* to make certain deductions—although operating in ephemerality. Being aware of continental drift and our changing earth and environment, backed by sound scientific evidence and mathematics, is, however, a Physical sphere activity with future scientific benefit assisting our evolving epistemology.

Mathematics remains the most reliable method to create some certitude in our Physical sphere. Whether we measure time, distances or forces, it is appointed in units to aid our Physical sphere, to enable us to put a man on the moon, erect mega-structures, and create all the technology we now experience. Its foundations are in the objective reality of numbers (presences or absences) where if an object's presence relates to 1 and its absence to 0, and we then add another, we now have the number 2, and so on. In this lie the primitive foundations of how we developed our ability to approach an airport in London, or create structures like the new Beijing airport and the Kingdom tower in Riyadh, without it all colliding or collapsing. Continents may drift, and universe(s) expand, but numbers will remain our best security at the time and in a state of change with an evolving epistemology where time and change are infinite, and everything is ephemeral.

The Metaphysical then looms as the great unknown, pregnant with new ideas. It is neither in our Logical nor our Physical sphere, yet open to interchanges between all the spheres. It is simultaneously full of potential, affecting both physical realities and genetic potential. Prior to its creation, the *Star Wars* saga belonged to the Metaphysical, but once created, the characters now commonly entertain our Logical sphere, with perhaps (who knows) future value. A television set was unimaginable and belonged to the Metaphysical to both Socrates in his wisdom and Leonardo da Vinci in all his creativity.

We now have endless more potential to interchange and circulate ideas between these three spheres of perception. It is imperative to briefly define (perhaps redefine) the Metaphysical for purposes here. The ghostly image of the Metaphysical is not free from being based on an idea (absences of presences) and its effect and impact variably based on species, time, and their specific Logical and Physical sphere interactions. To the new-era human, the metaphysical potential is vastly different from that of a Neanderthal. As unknowns get altered with our new understanding, so too does the Metaphysical evolve continuously with the Physical and Logical spheres, all evolving in complexity. We can see how cultural or religious belief systems can, by deceptively employing the uncertainty of the Metaphysical to set a normative, impair our cognitive abilities. Belief systems may become a set practice or a ritual, even if merely existing as an idea in the Logical sphere. Such practices and rituals can remain established for prolonged periods, kept alive by their main benefactors and complex social structures. With their core beliefs, however, remaining metaphysical, they operate at the best in the Logical sphere, with yet a significant impact on an evolving epistemology and physical outcomes. Examples can be seen in extremes from religious beliefs to taking supplements with uncertain benefits. More will be said about cultural and belief systems in chapter 5.

## 2.6 Continuity, Idea Making, and Interconnection

We live in an era where interconnection has achieved levels never experienced before. With the Internet now taken for granted, our cognition and moral demands are then also setting higher expectations than ever before.

We are compelled to clarify what is implied using the word 'idea' in our argument here. Avoiding over-involvement with philosophical discussions on what an idea is, we define an idea as *an interactive concept of objects resultant from either genetics or thought, or unavoidably both.* Thoughts in turn, being the products of a somatic brain and a perceptive drive, are epigenetically determined and open to trial and error. Such thoughts and ideas are constantly tested against changing conditions in an interconnected physical world. The Internet is a recent expansion of our perceptive capabilities following a similar pattern of rhizomatic interconnections in a perceptive network.

We can now begin to create the first ingredients of a simple formula:

Metaphysical⇆{(ideas⇆belief⇆practices⇆rituals)⇆Logical sphere}→objective value becomes Physical sphere realities

As an example, the idea of little green men on Mars existed very insecurely in our Logical sphere as a fiction and as cartoon characters passed on by folklore. Before its more modern alien association, it can be assumed (who knows?) that it originated in folklore to describe various supernatural beings, and later reappeared in fairy tales and children's books as goblins. Chris Aubeck recorded several examples of the latter in nineteenth- and early twentieth-century literature. Rudyard Kipling had a 'little green man' in *Puck of Pook's Hill* from 1906. The whole concept possibly may have started with the hallucinogenic effect of a toxic plant or as a physical play on light affecting someone, or with a drunken storyteller for all we know. Various beliefs and disbeliefs may influence an empirically based science and even establish new concepts with significant social impact on our Logical sphere of reasoning and even affect the Physical (remember ovism). Evidently there are no little green men on Mars and the idea will most likely soon completely disappear into the Metaphysical. Should we find any evidence of microscopic life on Mars, or perhaps now more likely in the inter-galactic Trappist-1, it will become established in our Physical sphere and then deduced that vast numbers of planets out there may also harbor life in various forms, from green to pink. Currently all we can do is place extraterrestrial life and their potential phenotypes as a possibility in our Logical sphere. This logic, which may appear a bit far removed from our topic currently, may perhaps be based simply on the arrogance in assuming that a vast universe or even multiverse evolved with us as the most advanced perceptive beings, or deducted from evidence of what we know about how life evolved under conditions here on earth.

Strangely, we have some strong beliefs without any physical proof, and practice some rituals (a meeting of the Flat Earth Society) without any objective reason except perhaps to have an impact on social structures. Beliefs, rituals, and their practices will also be discussed in view of their manipulative effect on both the Physical sphere and morality in later chapters.

Our genesis so far presents us with:

1. A principled yet pliable and perceptive genetic code branching out to assess the world in a network escalating in complexity while advancing cognition.
2. Constant change employing a pliable blueprint of RNA, DNA, and epigenome for protein production and electrochemical means to interconnect an organism, *a*, with a transfiguring environment, *b*.
3. Diverse methods of perception operating as pliable means to

formulate and interconnect ideas of *a* about *b* interacting between a Logical sphere, Physical sphere of reasoning, and the Metaphysical.

4. These ideas (whether seen objectively as genes or subjectively as cognition) have the key attributes of being perpetually progressive, interconnected, subject to valuation, vulnerable to external exchanges, and receptive to novel ideas.
5. Ideas may remain with uncertainty in the Logical sphere for various lengths with weaker or stronger presence, or return to the Metaphysical. Some genetic uncertainties with seemingly no substance may hover in the Logical or Physical sphere (schizophrenia on a genetic level) and fairies (on a cognitive level).
6. A Physical sphere of reasoning is secured by inarguable pragmatic physical evidence. The Physical sphere can also be influenced by false beliefs or pseudoscience that may again return to either the Metaphysical or Logical sphere at any time.
7. Concepts in the Physical sphere continuously circulate: a star turning into a supernova, Flat Earth concept, the epigenetic shift of black moths in the soot of London returning to a regular white color after the industrial coal-burning era—Physical sphere ⇆ Metaphysical.
8. Ratiocinated now is the vital evolutionary need for *impermanence, change, pliancy, transgenerational interconnection, perception, and uncertainty*.
9. We can also see the importance of the Metaphysical to stimulate a rhizomatic evolution based on interchanges to drive an interconnected evolutionary perception. Clearly, little argument is needed to see how an evolution set in a static environment, with a set theory of everything and no change or unknowns, is non-viable.
10. Using deductive inference, we can also conclude that if there were no unknowns or uncertainties (metaphysical), evolution would have no need to be perceptive, and life without inter-

connecting, and some perceptive means to do so, cannot be defined as life. Likewise (and this is now the good news) evolution, set in uncertainty and perceptive of change, has no limits and morality can also evolve.

11. The Metaphysical, continuously interacting with the Physical and Logical spheres, is also not exempt from change—the unknowns evolve as the other spheres evolve.
12. As explained, all spheres of perception are dependent on time, who, when, what you are, and where you are, as 'ephemeral reality.' This is now also in support of the Heisenberg principle, where a particle's position and momentum cannot be measured simultaneously because the measurement itself changes its location and/or momentum all the time (in different dimensions), while the 'measurer' also changes.
13. Cognition is shown by science to be a product of physical and chemical entities interacting and interconnecting. In its advanced form cognition can formulate concepts such as numbers and mathematics to improve understanding of the world around it. Each such is an interactive particle connecting with others in increasing complexity and escalating levels of awareness from ideas in a network of changing ideas. Our reasoning (and us) are as ephemeral and elusive as the change it is trying to capture.
14. We can now, without being accused of abstractionism, claim that the mind is as much an idea of the body as the body is of the mind.
15. This idea-making process seated in all three spheres of existence has a distinct pattern of testing, valuing and improving its world in a progressive epigenetic perceptive network. In a phylogenetic manner it advances complexity in interconnection and understanding.

We now enter a new era in post-reductionist evolutionary biology armed with:

> Elementary particles→proteins→cells→organelles→organisms→human brains→global communication (Internet and technology)→interconnected progressive cognition→more functional ideas→a better world with escalating moral demands

We can now formulize evolution (Evo) as a product of cognition and morality:

$$\text{Evo}=\text{C(mo)}\sum(\infty\Delta a\approx\infty\Delta b)$$

*where C=cognition and mo=morality

Note: Evolutionary morality is used here instead of merely evolution (Evo). The reason for this I hope is clear by now—we simply cannot explain life in its interconnected complexity without seeing it as *a principled* decorum with some *perceptive* means.

* * *

The Metaphysical is now a constant supply of contemporary ideas to a Logical sphere, and when these are attuned to a physical world, we obtain access to the clear pragmatic Physical sphere of reasoning. Accordingly, new ideas expand and become more interconnected and complex than before, and when greater stability is achieved it is also associated with greater reactivity. We can see how *objects*, *recognition systems*, *cognition*, and *interconnection* are simultaneously essential for evolution to function and that it is self-enhancing as it evolves. Also vital to such an evolution are pliant and adaptable recognition systems (valuation), interaction (value), change and motion with a strong interdependency, *and* reliance on a *sound ethic* to ensure its stability.

Value is now based on a non-discriminatory interconnected, pliable cognition where everything is measured and as important as the next as transforming keys in a search for workable ideas to a better world, all contained in the same network. Self-preservation and respect now become a duty as part of a larger whole rather than a solipsistic means of survival. If we do not see a form of altruism and the demand to act ethically in this interdependency, we will have to redefine the meaning of the word *ethic*.

From where we now stand with an interchange of ideas between our spheres of reasoning, harboring ideas from the Big Bang, quarks, and quantum mechanics, to a genetic drive with infinite idea-testing, we need to constantly remind ourselves that everything is continuously interacting, interconnected and changing—in need of a reliable ethic. Simultaneously, everything, including the potential of the Metaphysical, is constantly evolving. Our first failure lies in trying to manifest fixed ideas and theories that are not falsifiable, to fix a normative, or an unbudging Physical sphere in life, ignoring the constant change and interconnection that is vital to life and an evolutionary cognition. Our second demerit is trying to sway such contrivances for personal gain.

Enormous tolerance set in a sound morality is needed to drive a perceptive evolution, the keys to our existence as civilized cognitive beings. Any evolution set in a fixed normative or egocentricism detached from a physical world, using simple deductive reasoning, simply cannot still be called evolution, or moral. Subsequently neither can invariant knowledge be authentic knowledge. Simultaneously, any evolution not resulting in a progressive cognition in a principled epistemology cannot relate to these requirements, and these issues, as we now know, cannot be ignored.

Acceptance of these unavoidable inferences inevitably leads us to the need for trustworthy interconnection with pragmatic responses and correct behavior within a network. This is so that a progressive evolution can run its course, and how can we realistically turn our backs on our duty here? Besides pliancy, we can now also see the vital need for morality to progress as a civilized society. It is here in science that I believe we find clues to what morality is. It is also here where we can rid our epistemology of manipulative and immoral actions (common today) so that a truthful knowledge and science can evolve our morality.

With our current understanding in both physics and natural science we can empirically conclude that a fixed objective world or universe without constant change as a stimulant is simply not possible and cannot explain our existence as perceptive beings. Change innervated by an evolution stimulated by unknowns is therefore vital to life in an expansive cognition.

Some may still, on very insecure moral and intellectual ground now, argue that temporary manipulated and 'false truths' (as a method to perhaps create pliancy or seen as a survival skill to outmaneuver others) may be accommodated in such an evolution. We may argue our specious presence (see explanation below*), reality, and perhaps foolishly progress material fixations to nihilism. The latter now clearly appear unrealistic in a dynamic and progressive evolution searching for pragmatic truths facing harsh objective realities in a Physical sphere. For a sober and ethically orchestrated cognition to conduct apt but pliable responses to continuously evolve, change, uncertainty, and dependency are therefore imperative.

Consequently, future sociopolitical concerns and options will also have to revolve around this contemporary understanding; these will be considered next in the context of our current argument.

* * *

* The term "specious present" was introduced to philosophy and psychology by William James, in his influential *Principles of Psychology* (1890). It can be explained as follows:

> It is the view that we experience the present moment as nonpunctate, as having some short but nonzero duration. It can be illustrated by comparing our experience of the 'now' or present moment with the way the present is represented on a timeline. Both mathematically and physically, the present can be represented by a single point on a timeline separating past from future, moving along the line from the past towards the future. Such a present moment has no

duration. In contrast, the temporal character (physical presence) of our experience at least prima facie seems to span some duration, one that might range from as short as several hundred milliseconds to [James's proposal] as long as 12 seconds or more.

* * *

We can now categorize some options as follows:

- Continue to let the current archaic economic construct and false beliefs serve as core determinants of our vulnerable evolutionary epistemology, with change and time *not* waiting.
- Force laws and a normative through sociopolitical structures and their skirmishes. The forceful application of uncertain truths with no pliancy by an elect few powerful figures, mostly motivated by profits or self-interest, has been proved to fail historically and cause much suffering in such a disregarding drive, besides all the delay.
- Force a 'new' ethic that envelops the above and is simply another form of primitive grapple.
- Argue that an ethic cannot be practiced without dogma and forceful rule and surrender to this outcome as our only option. The immediate concern is that such an imperfect and obtuse system will always eventually be controlled by a few powerful figures and affected by cunning manipulation. Besides lacking morality, being unstable and resulting in needless strife, it will manipulate truths and delay our only hope and means to a better world in a progressive cognition avoiding a pseudoscience as a base.
- Consider harmonious and ethical coexistence and pragmatic truths as not possible or as a necessary evolutionary recruit—anarchy, ongoing strife, and barbarism are not only unacceptable to our current moral and cognitive status but detrimental to its ongoing evolution. This has escalated to a new level with

the responsibility of a science now carried to levels where it can cause mass eradication if not properly applied.

- *Or* we now have the choice to accept ourselves as interconnected, perceptive, and moral beings interconnected to the environment and part of an inescapable, interactive principled evolving universal ethic (cognitive drive). Such a demand creates better understanding and in turn also elevates our cognizance and morality to a never-before-seen universal level.

Coexistence with a universal ethic in a pliable cognition and set in a felicitous epistemology, is, besides its appeal, no longer idealistic but essential, in what is rapidly becoming an increasingly precarious-looking future—should we not change our current systems. And change we can, and *must* as will become clear—making this ethic also immune now against accusations of being quixotic.

Any dogma is and can never be an option in a quest for a universal ethic as it will always remain troubled and challenged by the question of whose dogma is controlled by whom, and open to power struggles, and inevitably conflict and violent confrontation—no matter who is involved: the Church, the powerful, the rich, the scientist, the politician.

* * *

Perhaps we can now in summary expand on Descartes' famous quote, "I think, therefore I am," to state: I exist in my Physical sphere of reasoning in a constant state of exchange with my Logical sphere, interacting with the Metaphysical while evolving my perception and understanding, interconnected to a changing world and universe.

We can now formulate a constant state of exchange between the Metaphysical and Physical as follows:

> Physical (object) and perceptions of objects ⇄ Logical sphere ⇄ Metaphysical, with interchangeable combinations explained in the next section (LSR⇄PSR⇄M)

We may be alarmed by the impermanence of fixed ideas in the above but also sense a hint of this necessity to drive an even more progressive, truthful, meaningful, and pliant Physical sphere toward a reliable epistemology.

We can claim that cognition is both intrinsic and extrinsic in a pliant and perceptive evolution, actively progressing by formulating and testing ideas around changing ideas.

* * *

Returning now briefly to our primitive slime mold, if it moves away from a light beam it perceptively does so, genetically primed to avoid desiccation. It does so by means of proteinaceous receptors using as signals chemical and protein molecules, initiated by messenger RNA to cause movement away from photons. The slime mold has no means of semantics, organic chemistry, physics, or physiology. Such ideas remain metaphysical from its perspective. Only the light beam is reality and part of its diminutive physical world, small 'epistemology,' and Physical sphere, with all else obscure. It, however, still 'knows' and is aware of light, and on a precursory scale, exhibits a primitive cognition.

It has become less eccentric to claim that the slime mold is ethically bound to avoid light. Now seen as part of an interconnected network within a complex perceptive drive, it is primed to do so. Note, not merely for self-serving survival and 'adapt or die' reasons, but with a transgenerational genetic responsibility. A true universal ethic would cross species, time, and spheres, and would remain as a pragmatic idea to serve in a combined effort. It would operate without discrimination and radiate freely from its origins in a moral interchange of workable ideas between all spheres as discussed—without prejudice or biases. The more aware, interconnected, and ethical this perceptive mechanism is, the better it can preserve itself and overcome ignorance while formulating new workable ideas as part of a united perceptive network.

With the environment vitally intertwined in such an evolving cognition, a search for an improved world now becomes goal directed.

## 2.7 Interactive Spheres of Cognition and a Universal Ethic

We may next ask: Why do we not merely propose that a **universal ethic** drives interchanges between the Logical and Physical spheres and the Metaphysical? Clearly, this would again place the burden on the nebulous metaphysical and externalisms to enter as a dictator of values and leave cognition standing insecure and vulnerable to manipulation. It also would confront all the criticism rightfully due to external moralism, as well as face the multiple dangers of an uncompromising power struggle to force, direct and dominate such a normative. On the other hand, moral particularism, where a perceptive morality is constantly evolving intrinsically, *together* with the complexities and demands of changing life, is simultaneously more comprehensive and pragmatic. Soberly conducted as an infinite interchange of flexible ideas, it confronts the demands set by constant change in an evolving universe. Such a morality can confidently confront unnecessary criticism and defend itself against the primary moral crime of egocentricism, and also importantly so against *reductionism*. This is similar to the new understanding of our evolution as now revealed.

Seeing the universe as a classical Kantian *Ding an sich*, or 'a thing as it is in itself,' and perhaps reflecting on how our ancestors saw the night sky 500 years ago, we can assume they did not relate to it the way we do now, and yet they pretty much saw the same night sky (allowing for a few minute intra-galactical movements and events hard to notice for the untrained naked eye). We can sense the impermanence of both our ideas and the objects we formulate these concepts around as affected by time and change. Classically, this egoistic outlook of the universe was challenged by the Copernican heliocentric solar system in 1507 and sadly instantly evoked enormous opposition from the Church, enforcing its dogma based on equivocality in the metaphysical. Had we perhaps had access to a more realistic, pliable, and pragmatic ethic to support sober ideas and social systems, many lives could have been saved over the centuries (including that of Copernicus himself) and scientific advances perhaps would have come more quickly. It took, however, many more years of 'sleepwalking' before even a Copernican universe became accepted by the average human and took us to where we are today. Now, when we look up at the night sky, we see countless galaxies and in astonishment with new understanding can relate to our diminutive place in our Milky Way. In awe, we accept but still struggle to comprehend the enormity of the distances in many light-years between celestial bodies in the context of our own ephemeral presence here. We can also now slowly begin to sense, aided by a progressive science, how our galaxy as one of billions is *interacting* and evolving in a perpetual evolving cognition. Yet, we still face similar obstacles to Copernicus, now over 300 years later, when it comes to our *moral* maturity.

In his *Critique of Pure Reason* (first edition 1781) the German philosopher Immanuel Kant stated:

> Everything, every representation even, in so far as we are conscious of it, may be entitled object. But it is a question for deeper enquiry what the word 'object' ought to signify in respect of appearances when these are viewed not in so far as they are (as representations) objects, but only in so far as they stand for an object.

We can see how our interpretation of object is subject to an idea backed by experience within an evolving epistemology, and may even superficially appear Kantian in origin. Having now firmly entered a post-Kantian era where it has become almost unimaginable to agree with him on his stance on animal morality, we can also not be pardoned from ignorance if we are not aware of and allow for interpretation of the changes in the object. The stars, the moon, the sun and universe all undergo constant evolutionary change, objectively but also based on unfolding ideas in our Physical and Logical spheres of reasoning. With change as an interminable, and with ephemeral interpretations and ideas about them continuously evolving, it stays an infinite process until there is no more change—a bit hard to imagine. Without change there will be no unknowns, no need for perceptive evolution or any 'us,' or as discussed, an evolutionary established life. Such an assured conclusion is made using simple deductive inference combined with current scientific knowledge.

Such change is not merely based on 'experience' as Kant suggested, exposed to an undeveloped science at the time, but based on a *perceptive* evolutionary drive. Vitally, so we now know, the objects we study are also constantly evolving and changing as a prerequisite for a progressive cognition and life. Deductively, our Physical sphere is dependent as mentioned on what, who, where, and when we are there, and consequently, on what we already know (as what we are, under the above criteria) about the persistently changing and interacting objects we perceive, when we are there. We should again sense and reemphasize, not only the interconnectivity of it all, but also the constant change in perception on all levels. In constant flux and exchange and dependent on this united metamorphosis and its complex interconnections, our spheres of perception continuously evolve and escalate our understanding. There is not much scope for overwhelming dogma, egocentrism, or fixed ideas in such an overwhelming scheme of infinite ideas and ongoing change—beyond their temporary pragmatic and ethical value to this evolving intricacy.

* * *

We can now also present adroit ideas to our malleable Physical sphere and circulate them between the three spheres of perception in six-fold orders—calculated as 3x2x1=6. This gives us various orders of unfolding and disseminating innovative ideas.

An example of each order is randomly given below to entertain this prospect:

1. **Physical sphere** ⇌ Metaphysical ⇌ *Logical sphere*: Estab-

lished knowledge returning to the Metaphysical with the future potential of returning to the Logical sphere.

A medical example can be used here (one of many), in the short-held idea of the value of increasing aromatic-amino acids in the diet as an aid in treating Alzheimer patients. This was arrived at based on measuring lower levels in the Alzheimer's-affected brains compared to the non-affected brain. After a period of raising these dietary proteins in the Alzheimer patients and with no benefits seen, the treatment for most current clinicians now belongs to the Metaphysical, but the concept remains as an idea in the Logical sphere of researchers in this field, perhaps to return to the Metaphysical or again to the Physical. Interestingly on this topic, there was at one point even a 'Brain Diet' for dogs based on such anecdotal evidence. Quickly rushed onto the veterinary market at great cost to consumers as an 'aid' in treating cognitive dysfunction in elderly dogs (well removed from its pathophysiology in Alzheimer's disease), driven by those keen to sell, it was soon to be withdrawn into the Metaphysical/Logical sphere.

2) *Logical sphere* ⇌ Metaphysical ⇌ **Physical sphere**: A known idea under consideration in the Logical sphere of reasoning returning to the Metaphysical for a period before being revived for pragmatic application.

Examples here include the rediscovery of the importance and function of the previously overlooked glial support in the brain. Historically overlooked as an inactive support structure for neurons, more recently we understand that glia assists in neurogenesis and transmission of signals in the brain, as discussed before. Another example (of many) here is the contribution of the spirochete *Helicobacter pylori*, living in the stomach lining and playing a role in stomach ulcers. In 1875 it was with much unclarity noted in the stomach lining by German researchers, with uncertainty about what role it played and even where to classify it (Logical sphere) and soon it was completely forgotten (Metaphysical). Barry Marshall and Robin Warren's rediscovery, gaining them a Nobel prize in 2005 for their work, returned this bacterium to our Physical sphere as the basis of a now recognized treatment for stomach ulcers, albeit with ongoing open questions for new research.

3) *Logical sphere* ⇌ **Physical sphere** ⇌ Metaphysical: Prototype flow, in science and pharmaceuticals, of ideas under scrutiny and after exposure to EBR as discussed, becoming sustainable in our Physical sphere but with time potentially returning to the Metaphysical and replaced by new ideas.

   Many examples exist here, including medical techniques, drugs, and practices, as already discussed (ovism, etc.).

4) **Physical sphere** ⇌ *Logical sphere* ⇌ Metaphysical: Workable contemporary ideas under critique returning to the Logical sphere of reason with uncertainty and eventually to the Metaphysical.

   Again, many examples exist in medicine here; another may be seen currently in physics with the changing opinion on a fixed speed of light.

5) Metaphysical ⇌ **Physical sphere** ⇌ *Logical sphere*: Sudden appearance of novel ideas needing no further proof to enter the Physical sphere of our cognitive life (for example, discovery of a new species or planet) with futuristic potential to interchange with the Logical sphere ('Eureka!' moments).

6) Metaphysical ⇌ *Logical sphere* ⇌ **Physical sphere**: Where unknowns (M) appear with uncertainty in the Logical sphere of reasoning with the potential to occupy our pragmatic Physical sphere of reason.

   An example is when a new star on closer scrutiny turns out

to be a galaxy.

These examples can perhaps be juggled and articulated in different ways, but the main aim here is to introduce spherical interchange of reasoning as a method to circulate ideas, *avoiding reductionism and attachment.* We can gather from the above that ideas circulate in three spheres of altering reliability but in a concordant perception (unless manipulated or afflicted by disease). We can clearly see the need for a pliable valuation system, constantly interconnecting reliable ideas while everything is constantly and freely changing and evolving; and the inadequacy of any *set* blueprint (a fixed sphere) isolated from the requirements of the unfathomable perspicacious cognition in which all these spheres interact. We can also see how, when inhibiting, manipulating, or suppressing reliable interchanges between these spheres, it can affect pragmatic outcomes and the overall intelligence of such an open design. Modern science presented us with a world interconnected with measurable objects (material world) seen as consisting of interactive atoms and molecules; now we also see it as an interconnected perceptive evolving sagacity—we are obliged to address the equally progressive moral demands that come with this realization.

In science, we now have many more opportunities for a progressive cognition to reliably interact and soberly evolve our understanding and an urgent demand to protect the trustworthiness of these interconnections. The traditional philosophical metaphysical is no longer seen as a 'dark ocean' but as new potential, helping to evolve a dependable and principled epistemology.

Kant's conveyance of morality to the noumenal world, a world where objects and events are beyond sense and only human beings qualify because of their ability to think and rationalize, now appears unrealistic and extrinsic. A world separate and distinct from, and *not* interlinked to a world of objects or phenomena, now truly appears unworkable, with sense (perception) having its origins and substance in a vitally interconnected world of interactive particles and changing phenomena. Kant's idea of two worlds existing separately—the world of sensible objects (phenomena) and a noumenal world—is also untenable and can be claimed as perhaps lacking morality in view of our argument here. Morality now becomes an inseparable part, vital in its role of interlinking pliant evolving phenomena in an infinite cognition that *evolves itself simultaneously, both intrinsically and extrinsically.*

* * *

If we now turn again to our basic building blocks of life, the nucleic acids DNA and RNA, and look at the way these principled nucleotides are employed, we can see an *a priori* similarity in our spheres of perception (we scrutinize this more closely in the next section).

The potential to exchange reliable and innovative ideas and evolve knowledge has now become tremendously larger. It is also more attractive and comprehensive than under a conservative 'conniving' natural selection where a solipsistic gene was fighting for survival, armed with an obtuse blueprint in an isolated Physical sphere. Simultaneously, this comes with increased moral responsibility and reemphasizes the importance of nurturing a sober 'ethically' evolving perception as our most important mission. With such a moral responsibility now a pressing phenomenon, we can no longer afford to ascribe to an implausible insensitive world set in uncertain beliefs, and turn our back on the growing incidence of mental health issues and corruption in our society. Or, in its worst form, force a set normative, based on personalized world-views and power, from such equivocality.

There is a similar privation evident in a Kantian noumenal world and our traditional interpretation of natural selection. Under a previously misconstrued public view of a materialist natural selection with emphasis on survival of the fittest, a poor ethic emerged and could easily become manifest, allowing for self-centeredness, greed, corruption, manipulative behavior, poverty, and generally a disregard for others and the environment. A new post-reductionist evolution is also in contrast with natural selection's previously proposed core drive of survival and reproduction, now with evolution understood as a perceptive, all-regarding, interactive evolving *network*. This is now a much more benevolent world we are evolving into, deserving of much more respect for all its containments. A world in which we can all aspire to morally and perceptively co-evolve, rather than merely survive. With time and change driving a preceptive evolution, the commodious design and purpose of life itself also become less despotic than under previous concepts.

## 2.8 The Genetic Code—a Moral Code

We can now recognize the ineptitude of a morality formulating its core values around a vacuous unperceptive world, or a world entirely based on an atomistic and 'self'-centered natural selection. Assuredly, we now understand how such disjointed and mechanistic worlds will result in no more than a battlefield of overinflated egos arguing around vacuous beliefs while searching for unfounded security in the metaphysical. All are dismally failing in meeting the pragmatic requirements of an interconnected evolution. Such an evolution is a much more progressive and gracious network, simultaneously perceptive and communicating principled ideas around changing matter.

We need to briefly reflect on the recognized basic mechanisms (in our Physical sphere) of the genetic drive here:

Classically, we now accept four nucleotides (RNA uses an extra one, making it *five*) which harmoniously alter sequence to produce numerous different genes and produce a variety of proteins. These proteins communicate the messages of life as we know it. Besides fulfilling many actions as specified by genes, these offspring proteins also act as enzymes and communicate bonding between molecules. One of their prize achievements in their search for a better world is **cell signaling**—the way cells interface with other cells and the outside world.

Cells *communicate*. We can now claim, without being accused of eccentricity, that cells 'message each other' using various methods. They use distance communication through hormones (hormones likewise are proteins produced from strands of RNA) referred to as **endo-**

**crine signaling**. Neighboring cells have many ways to communicate using ligand synapses and other means referred to as **paracrine signaling**. A *ligand* is defined as the binding of two ionic molecules, normally between a larger and smaller one. They 'chat' to themselves with **autocrine signaling** where they produce (from strands of RNA) their own internal protein ligands to create unique identities, even as part of the same organ. This bears an uncanny similarity to the way personal views stimulate social interaction, and communication is seen here on intracellular and intercellular levels reaching into the external world. We have already covered much of the mechanisms of this actively interconnected biochemical world in the preface to this book.

In brief, to recap, we know that cell signaling is part of a communication process that governs all basic activities of cells and coordinates all cellular actions and reactions. This process escalates up to eventually sustain higher forms of cognition such as the human brain. The vital ability of cells to be perceptive, to interconnect, correctly respond to neighboring cells and communicate changing needs to this network, forms the very basis of development, tissue repair, immune response as well as normal homeostasis. In turn, any error in communication and information processing is responsible for cellular damage and numerous ailments such as allergies, autoimmunity, cancer, diabetes—culminating in the death of the cell or cells. By unraveling this microscopic world, and its paramount need for interconnection, cooperation, and principled conduct (doing what is right), there is much potential to prevent decrepitude and adjust to the demands of functioning adequately in a much larger living network.

The five nucleotides mentioned above are hooked together in different orders to form strands of DNA or RNA while also modified by its epigenome. This epigenome creates the ability to modify or switch certain parts of the genome on or off as directed by changes in the environment. From this platform is orchestrated the production of the numerous proteins required for cell signaling and the metabolic and physiological reactions responsible for the expansion and diversity of all life—impossible to progress in isolation without stimuli and means to perceive such input. These nucleotides are composed of (surprisingly plain perhaps) a five-carbon sugar, a nitrogenous base, and a phosphate group. Such decorum serves as the building blocks of RNA and DNA, then also production of the proteins that make our slime mold move away from strong light, or transmit signals between the synapses and neurons in the mammalian brain.

These nucleotides are further classified as the purines adenine (A) and guanine (G) and the pyrimidines cytosine (C), uracil (U), and thymine (T). The genetic code for the variety of life and our ability to perceive it is determined by these seemingly simple base-pairing codes in different orders, directed by their epigenome both *a priori* and *a posteriori*. This is an unembellished and pragmatic method that has worked so far in matching the need for pliancy and diversity in a trenchant perception, compelled to react appropriately to continuous change on various levels. All the complexity of life we witness around us is based on these base-pairing rules between a few nucleotides and their pliable arrangements affected by the epigenome. Adenine (A) combines with thymine (T), and guanine (G) with cytosine (C), by means of covalent bonds; RNA uses uracil (U) instead of thymine (T) as a pyrimidine. A bond is called *covalent* where electrons between atoms are shared. With a fairly basic description of these building blocks relayed here, we must also accept that from this objective 'simplicity' eventuated the human brain and evolve all our thoughts. Above all, we must realize that without a means to be perceptive and reactive to evolving change, these building blocks could not function. It is in this progressive change and boundless perception between matter that I believe we should search for sagacity and evolve our morality and not by merely forming attachments to its fixed objective values.

After discussions here, we can now also relate to nucleotides and base-pairing rules as being set in a more pliable Physical sphere where outcomes in the Logical sphere evolve to match changing demands while simultaneously evolving the Metaphysical. This is not a fixed blueprint of life, as previously seen, but an existing workable and pliable combination that is obliged to interact with escalating complexity and demands in a perceptive network. In the fields of astrobiology, it even becomes more plausible and attractive to entertain (Logical sphere) an array of other lifeforms (perhaps more benign and advanced in nature than current fictional presentations of aggressive invaders) existing somewhere out there, using different base-pairing combinations. Likewise, our own genetic potential now appears boundless and to be evolving morally rather than in disarray left to its own devices. This releases us from the dead-end reached when entrapped in the previous reductionist version of a set DNA blueprint under a ruthless natural selection, leaving it entirely to us to decide its destiny.

With current understanding, covalent bonds between the sugar of one nucleotide and the phosphate of the next, cell signaling, and ligands are now pragmatic Physical sphere activities. Such affirmations managed to sustain a growing interconnection of 'ideas' to expand from simple strands of RNA to our current perceptive means to map our own genome. We can now also accept in our Physical sphere of reasoning how simple principled sequence-changes in these versatile idea-forming nucleotides result in different gene and protein expressions, forming the variation of life experienced on earth. We can also newly consider why some of these nucleotide sequences may form non-coding genes with no apparent productive value (as discussed before, similar to neurodevelopment). It is becoming increasingly plain that these 'non-coders' may serve as important co-workers and act as a reserve for the plasticity needed, also with *potential* to form new life-altering proteins in anticipation of unknowns. Perhaps not surprising, after our exposition here, they are proving to be much more significant as active contributors in facing escalating demands of an evolving metaphysical, and who knows what they may still reveal. Progressing in synchrony with an interconnected perception, they formulate pragmatic and progressive innovative ideas while confronting change around them—constantly reiterating the obsequious demand for harmonious interaction on all levels of existence as part of a united perceptive drive.

We have already mentioned how one of the central dogmas in biology has changed, with acceptance of RNA now considered as the original orchestrator of DNA and life, with DNA synthesized by reverse transcriptase enzymes (proteins) from strands of RNA. Yet, these are not random events with no ethic, value, or rules, but are perspicaciously driven to expand, interconnect, and perceive. Reproduction and survival have now become mere implements to sustain an infinite and rhizomatic 'idea-forming' process. Recent research in epigenetics now also suggests that what we (in a needless derogatory manner) previously referred to as 'junk DNA' makes up 98 percent of the human genome and has the ability to be 'mobile' with transgenerational genomic impact. The set DNA (the exons) encode only a striking 1–2 percent of mammalian and human DNA, with opinions on this value constantly changing.

What a turnaround, what potential and what an awakening! With some relief we can now accept that a much more versatile DNA than the earlier 'blueprint' can constantly cut and paste itself, with a reserve to back it up in response to equally active and diverse environmental and cellular challenges. With a discerning plasticity and long-term aim, this now replaces the earlier reductionist view of a DNA set in its inhibiting short-sighted blueprint. This is also where we should not attempt to play God by manipulating outcomes. Central dogma, insular concerns (egocentricism), or a set normative can no more be seen as part of a viable evolution (and cannot be), not even on the DNA level, following a practical and pliable, universal, all-regarding 'moral particularism.' Survival as a main motive is now eminently overshadowed by a much more sagacious perceptive drive to advance a progressive coalescent cognition and morality.

It is important to note that a few sequence-changes (with potential for many more) in these relatively simple nucleotides persistently strive to perceive and harmonize themselves with changes in the environment. Similar sequence-changes occur between the Physical sphere ⇌ Logical sphere ⇌ Metaphysical, and continuously change concepts with a progressionist perception of objects and ideas about them. With both concepts (nucleotides and perception) interchanging and malleable, sequence-changes may undergo ongoing presence (get switched on) and disappearances (switched off) in the Physical sphere on all levels of existence (genetic and cognitive) transgenerationally. This self-enhancing process appears to be infinite while we evolve in an interactive constantly changing universe, and how can a universe not be expansive in a progressive cognition? Kant's insensible world and *Ding an sich* is now inseparable, interconnected, and interdependent, and so are we to all this infinite sagacity.

* * *

Gathering from the above, we can now postulate:

- Metaphysical dogma enforced by an imperceptible deity on a Physical sphere, *M→PSR*, will present a restrictive archaic view of a pre-Copernican universe.
- A set Physical sphere overshadowing and dictating the Metaphysical, *PSR→M*, will also be restrictive and immoral as a mere reversal of authority and insipid to a rhizomatic cognition in search of workable truths.
- Our current-day trend of a lagging morality standing alone while attempts are made to secure a Physical sphere based on an embattled Logical sphere open to manipulation, *LSR→PSR*, is also restrictive. This method further slurs a confined cognition and creates a disparaging ignorance of the Metaphysical.

Confident defenders of the above impracticable approaches, mostly subject to personal beliefs or self-benefit, consequently feel threatened by and strongly object to more liberal prospects. We may perhaps be forgiven for some ignorance here, with our inborn primeval fear of the unknown, making reductionism lucrative to our ancestors, always looking for security in familiarity in a harsh world full of threats. The latter arbitrary behavior is responsible, not only for much strife historically, but also much unnecessary suffering today. With such a restrictive flux of ideas and knowledge open to manipulation and driven by fear, a becoming universal morality is also torn to shreds. Our current economic model is entrapped in such a restrictive mindset.

Even the superficially appearing liberating interchangeable *M⇌LSR⇌PSR* is still restrictive when compared to exchanges where ideas move multidimensionally between all spheres, on all levels bowing to a universal ethic and cognition.

We can now formulate this concept as:

$$Ev(mo)=\sum\{\infty\Delta a(\text{Metaphysical}\rightleftharpoons \text{LSR}\rightleftharpoons \text{PSR})\ x6 \approx \infty\Delta b(\text{Metaphysical}\rightleftharpoons \text{LSR}\rightleftharpoons \text{PSR})\ x6\}$$

Armed with nucleotides as pliant and perceptive building blocks, *a* can capaciously respond to changes in *b*. Perceptive in all spheres—Physical (PSR) and Logical (LSR) confronting the Metaphysical (M)—and by constantly interchanging, adjusting, and harmonizing pragmatic ideas, we can now evolve an equanimous all-regarding cognition and morality set in a trustworthy epistemology with endless potential.

Circulating our ideas between these spheres, we can also proximate our thoughts closer to the morality that exists between matter, instead of being *attached* to the matter itself. This will give us a better ability to preserve, nurture, and respect a universal harmonious cognitive wellbeing and apply goal-directed universal healthcare practices for all living systems. The latter may have seemed grandiose or idealistic even a decade ago. Now, with new understanding in the sciences revealing similar interconnections, pliability and interdependency on all levels of nature, the importance of nurturing a healthy planet in a sober cognition has become *imperative*. An interconnected environment and health can no longer be seen as separate issues. Neither can it be brushed off as a disillusioned ideal left to a few activists or unconventional thinkers. It now has become the core focus in planning a joint, more benevolent future, where the Earth hurts as much as each of us when exposed to greed, corruption, and ignorance.

* * *

When seeing our cognitive origins and limitations as an individualized struggle to survive confronted by extinction theories in the confines of a harsh and competitive environment, or alternatively clinging onto false beliefs, we can do no more than imprison a rather obtuse cognition under false dogma. This now-historical view of life as a fortuitous affair driven to reproduce, adapt or die is not only responsible for creating an unnecessary malignant and aimless view of life, but simply cannot support the diversity of the progressionist perceptive evolution we are all part of.

Now functioning in a more sovereign Physical sphere, and driven by a perceptive and sagacious evolution, we have secure and reliable spheres of perception where we can evolve and pliantly, yet confidently, coexist. We also have scope in a composed Logical sphere where we can trustingly tap ideas for a pragmatic perspicuous Physical sphere to evolve. Acknowledging change while constantly drawing on the vast potential of the Metaphysical, we can now do so without dogma and restriction but as guided by the wisdom of a truthful epistemology.

# 3

# Morality and Emotion in a Mobile DNA

*Shift your attention and your emotion shifts; shift your emotion and your attention shifts.*
Frederick Dodson

We come into this world with a somatic brain and a few genetically acquired reflexes to respond to our new world. Armed with this brain and six innate reflexes, as in the case of a newborn infant, we set out to perceive and formulate avant-garde concepts about an equally extraordinary world. We should also not forget the already mentioned transgenerationally attuned mobile genome as part of this armor here. Already in the womb, signals from the outside world can reach the brain, and as we mature, our brains experience the changing world around us subject to signals received by our senses and as orchestrated by our packaged, versatile DNA and RNA. This, then, is the basic genesis of how we formulate new propositions to create familiarity, express who we are, and grow to perceive this dynamic world—with a progressive perception and morality that is not all fixed and much more pliable and interconnected than we think.

Determined by our genes, this perceptive process is strongly influenced by environment, nutrition, health and education. As we develop, senses are our means to receive multifarious signals from our surroundings depending on what, who, when, where and how we are. Our senses expose and continuously link us and our brain to our habitat, interconnected and interacting in continuous change. The basic human (mammalian) senses to receive this perpetual change, traditionally are vision, touch, sound, smell, taste, vestibular (balance and movement), and perhaps we may include introspection here. In the end the value of our experiences, what we are and what we leave behind once our somatic brain ceases to function, is no more than the sum of the continuation of ideas that helped to improve the world for those left behind. This impact extends also to how we altered the environment, and more recently now as acknowledged by a growing number of scientists, how we nurtured our genome for future generations to experience their new worlds.

We can perhaps define a superlative perception as one filled with thoughts and ideas that improve the world and lives of other sentient beings in a shared habitat, assisting in shaping a better future for all (see *beauty* as analyzed in chapter 4, section 3). Such a magnificent mind would, without fixations, perplexities, or self-interest, operate on a higher moral level and in better understanding have ideas carried to future generations. In creating such a divine perception, well-attuned to the interconnected universality of all things, clarity in the Physical sphere of reasoning comes naturally and brings us closer to a universal wisdom and its progressive morality. This is now within reach for all of us, *if* we follow the code.

* * *

To explain what is implied here and avoid intangibility, we must momentarily turn back to our basic senses and what they are in the context of a perceptive evolution. We can again use as an example some migratory birds and their ability to detect the earth's magnetic field. These birds have already evolved the trans-genetic transmutable ability to sense and detect what we as humans have only more recently achieved through our advancing technology. Carried transgenerationally, these genes (with pliability) can be turned on and off according to environmental demand and change. Both methods are considered more progressive when understood as part of a perceptive genomic network. Each of such sensory abilities has obvious survivalist benefits to the individual or group with pragmatic value in connecting organisms to each other and the world. We should move beyond the primitivism of seeing such sensory means as mere chance-acquired endurances, there to profit the benefactor at the cost of others. These benefits only function and become meaningful when seen as operating in a transgenerational intercommunicative progressive network. Seen in isolation and as non-pliable and not adjustable to change, they appear meaningless and a mere means to an end. This end then also becomes as meaningless as the means—this reductionist view threatening to infiltrate and endanger the hard-earned evolutionary epistemology of an entire civilization.

In the historic version of evolutionary biology such an outlook was inadequately interpreted as 'naturally' selected benefits that randomly, and rather fortuitously for the organism, popped up with short-sighted value to a disenfranchised genome—improving its odds in a struggle to survive. Principally directed at the survival and reproduction benefits of its solitary gain, such an egocentric and reductionist outlook or belief hit an impeccable blind alley and can also make conniving, greed and war attractive. We can now confidently assume that evolution is much too insightful and resourceful to resort to such confinement of a universal perception. Our progressive cognition now also comes with much greater responsibility and more interconnected moral demands than previously realized.

We can begin to see how this previous restrictive view of evolution can even be accused of egocentrism, thus holding back a society and its growing epistemology that demand a much more progressive system for an interconnected world to grow to its full potential. In an interlinked progressive cognition, this is now seen as a rhizomatic all-regarding perceptive drive attuned by interactions between *a* (the observer) and *b* the observed, adjusting to the continuous change in each other. Simultaneously, the moral obligation and enormity of the task in such an all-regarding perceptive network now clearly seem more intelligible than under the earlier reductionist and more egocentric version, where focus was on individual outcomes and survival as the principal aims.

It also now seems impossible to disengage an evolutionary cognition from emotion and sense when understood as all part of a joint evolutionary perceptive drive with vested interests. Emotion is now also an inevitable and necessary product of evolution. Such an interconnected cognition evolves while simultaneously advancing both the moral demands and understanding of the mechanisms of this growing network. Various interactions and behaviors are now evident in science and seen in all things, from the atoms that combine to form organic matter to the escalation of cognition, from unicellular organisms to eukaryotes, to the eventually emergence of the human brain.

This radical disengagement from a previous disregarding and reductionist evolutionary concept is urgently needed to also help prevent the ongoing hacking of a universal morality exposing our evolving epistemology to a pseudoscience, false beliefs, and profiteering. Based now on a non-restrictive naturalism and without being bogged down by old-school reductionism and the narrow confines of what pays, we can begin to interpret how cognition, our senses, and emotions are interconnected, pliable and altruistic as 'concepts' concordant with the proteins and genes they evolve from. This represents something much higher and more principled than a mere struggle to survive and outmaneuver each other.

We can now see *emotions* as an evolutionary recruit taking over and playing a role where physiology and linguistics fail—working synergistically, both intrinsically and extrinsically, in enhancing the effect of vocabulary, as a necessary product of an escalating all-regarding evolution. Perhaps our sentiment expressed here can be seen as part of this process.

Improved (open) communication through senses and emotions (or whatever means) should not be reduced to merely increasing the chance of our survival but seen as the advancement of an interconnected burgeoning cognition that comes with increased moral demands as it expands. We may by now begin to appreciate how an interconnected progressive cognition existing in an all-regarding ethos and open, principled society may perhaps be the primary goal of evolution, and that it is not possible to function without full regard of these elements. A simple form of deductive inference can be used to further support this concept:

i. Without ascendency to interconnect in some form or another, there can be no interaction on either subatomic, molecular, cellular, genetic, or reproductive levels; no awareness and no sharing of ideas, genetic matter, or emotions; no evolution and no life.
ii. Should we, rather egoistically, aim to explain higher cognition as a genetic recruit to enhance self-preservation or as a rather adroit survival skill, or ascribe it to the realms of the metaphysical; and not disclose it as a mere instrument in the continuation of an endless exchange of ideas and understanding as part of an expansive interactive cognitive network with shared interests, it becomes both self-limiting and *devolutionary*. Compared now to a perceptive evolution based on its rhizomatic interconnections in a multidimensional open exchange of ideas that operates transgenerationally, the former now seems severely debilitating. Such an approach also links us to an infinite cognition, continuously evolving in complexity and understanding.

A progressive morality evolves hand in hand with an organic evolution, as we may see on an elementary level in even primitive lifeforms and in all the phylogenetic diversity of life and values we experience. It is now postulated here that the expansion of morality is based on an intimate and progressive interlinked perception, with emotion, sense and introspection as vital interconnected and active evolutionary recruits. Morality escalates, and as it does so, it advances cognition within an expansive network. With our understanding and perception thus evolving, so too does our morality; and reciprocally if our morality declines, so does our understanding.

Interconnection⇆(Morality⇆Cognition)

Inarguably, if we decline in morality we will lag in interconnection, and with decreased interconnection we will grow apathetic and careless of others and their needs and our world. Subsequently, as we gain better comprehension and more respect for our place in this interactive network, we can improve both our cognition, wisdom and morality. This vital spherical interaction between the spheres of reasoning (Physical, Logical, and Metaphysical), as presented here, is in a similar manner open and driven by evolving unknowns, as we have discussed.

We can deduce from this, then, that increased interconnection in a perceptive drive confronting more unknowns will self-enhance both its morality and obligations as it grows in complexity. As postulated, a primitive evolutionary morality already exists in strands of RNA searching for a means to communicate with their surroundings, with DNA as a pragmatic means and an idea that worked (chapter 1). There is no need in this text to further trace the social evolution of humankind from our animal origins, emerging as xenophobic tribes to today's complex web-connected 'global village'; many works in diverse fields cover this in depth.

Dissimilarities between people have become less obvious or socially significant compared to even as little as a century ago. Traveling to any small and remote town or village today, whether it be in Alaska, Asia, or somewhere tucked away in Africa, the most common query now relates to smartphone reception and Internet connectivity. Likewise, our application of moral rules has expanded from serving narrowly defined social groups with defense structures kicking in, to an expanding global community with joint concerns and universal moral and ethical demands.

Think of the Internet, smartphones, and social media today, and its inevitable effect on influencing many social values and norms. It has the amazing potential and ability to raise urgent and escalating global concerns about the environment or poverty and coerce apathetic governments and corporations guilty of promoting isolationism, *or* on a less advanced level it can get monopolized to spread destructive messages or to trap consumers. The ultimate outcome depends on the morality and the reliability of its data and information, freed from biases and greed. We now for the first time have the opportunity *to speed up our own evolution*, without harming or exploiting our environment, but through improved perception and better understanding of how everything interconnects.

## 3.1 Emotion and Connectionism

Waking up now to a new age of natural moralism and with the inevitable value of an advancing technology as support, we should, however, also be wary of becoming too attached to a connectionist model when it comes to robotics and Artificial Intelligence (AI). With a non-reductionist understanding of life, we must consider the possibility that the only thing connectionism (the current cornerstone for AI) may have superficially right is an understanding that there is an interconnection between neurons and neuron pathways in living matter. This is relevant to us here and even urgent, with the growing dependency now on stored data and threat of unemployment figures rising. Evidence for such concern is familiar; think of ATMs and how an expensive piece of hi-tech equipment or computer software can at times callously replace several workers, together with their emotional, personal and retirement needs. The impact of a developing hi-tech society is still predominantly focused on commercial short-term benefits, with little acknowledgment (or built-in values) to address its environmental and long-term social impact. This will, however, become a much more pressing issue and impossible to ignore in the near future. If we are not prepared to morally and pragmatically adapt the sharing of resources to the effect of climate change, and wisely use the benefits of modern technology, it could steer us to disparity and dissociation. Again, evolutionary biology can act as a guide here.

We must avoid being remembered as the followers of a reductionist era subservient to an Artificial Intelligence subsequently leading to their own demise, in a history that may unforgivingly be taken as barbaric by future generations. If we continue to ignore a new understanding of our interlinked *perceptive* evolution, being controlled by Artificial Intelligence, and indeed our own extinction, is something we may have to consider, rather than all the potential benefits that AI can offer a progressive civilization.

* * *

According to the traditional *connectionist*, a neural network consists of many units joined together in a pattern of connections. Based on such a dangerously reductionist model, these units, as part of a net, are segregated into three classes: input units, which receive information to be processed; output units where the results of the processing are found; and units in between, called hidden units. If a neural net were to model the whole human nervous system, the input units would be analogous to the sensory neurons, the output units to the motor neurons, and the hidden units rather equivocally to all other neurons.

In such an oversimplified and disjointed model, sensory input is then weighed and measured against objective value not at all representative of what occurs in an interconnected, progressive and perceptive evolution. Remember, a *pliant a* is interacting with a *changing b*—both persistently adjusting (harmonizing) to change with transgenerational input, all operating in unison. It is further suggested in the connectionist model that hidden units can then 'either enhance or reinforce a potential output.'

Besides the potential value, a multitude of limitations, questions and concerns arise from such a reductionist concept, and are important to us in our search for acceptable universal principles. Enveloped in this advancing technology, and the knowledge and data it could potentially contain, we need to now consider the following:

- How and with what do the 'hidden' units measure anything, compared to a complex and pliable somatic brain transgenerationally primed through its genes to continuously and with pliability adjust to new information, changing and malleable environments, and uncertainty, as part of a complex social network?
- With recent evidence of the plasticity of neuron pathways (with a 'mobile' DNA), a connectionist model will have an impossible task to adjust to the continuous flux and the 'constant weights and balances' of complex idea-forming in these hidden units,

as implemented by a perspicacious evolution aimed at adaptability and adjustability on numerous interactive levels. An enhanced version of the aforementioned Heisenberg principle and Gödelian inconsistency would also persistently pester any connectionist model (Gödelian inconsistency can be referenced elsewhere for those interested, as it is outside the scope of this book).

- The connectionist model will find it complex (perhaps impossible) to create the plasticity needed to constantly evolve and circulate ideas between the Physical, Logical, and Metaphysical spheres of existence in a complexity of diverse environments (continuously changing). The immediate problem for the connectionist model is that input and output is programmed and based on set statistical and historical values and conclusions based on statistical norms (expectations)—it is reductionist. *Nothing about life hints toward a conclusion or fixed values as it is driven by uncertainty as much as by pragmatic outcomes.*
- With even our Physical sphere of reasoning based on shaky objective values (varying as we have seen with who, what, where and when, and how you are), perceived and attuned with genetic variability and pliability, the connectionists' task now becomes not only unimaginably large but perhaps even impossible, reductionist, and in a sense, devolutionary.
- Allowing for the impact of emotions, and due to perpetual changes in the relation between an organism and its environment, any connectionist attempt to create a systemized model of an individual neural network would already have changed before completion of the model; as our perception now has the ability to also more rapidly evolve in more rapidly changing surroundings. Likewise, the organism and the environment are also in a state of perpetual change.

We can see how in a connectionist model we have an oversimplified model lacking plasticity, adaptability, and *emotion*, with its principal value utilitarian and subject to an evolving cognition. We can perhaps also sense the hollowness of expressing emotion toward any form of AI, perhaps with its latest update not agreeing with your world-views, or the foolishness of arguing with Siri. Inevitably an aid to our future progress, Artificial Intelligence will continuously struggle to match the transgenerational perceptive genetic plasticity, emotion and sagacity of a progressive living cosmos. With recent research in this area taking the first steps in brain-to-brain interfaces where brain signals can be transformed to affect mechanical movement, there is undeniable potential here. Limited to utilitarian value, we will soon encounter such developments entering the commercial markets. Undeniably, such advances may help paraplegics and others suffering from spinal-cord injuries to control machines with their thoughts and to bolster their ability to get around, but are still subject to a pliable infinite cognition and its morality. In addition, it is also not far-fetched to consider that we may in the future, with emerging nanotechnology, be able to manipulate these building blocks (DNA) to behave in a certain manner in curing disease. With such enormous potential benefits in healthcare we should, however, be very careful that without deeper understanding of how this principal perceptive network interconnects and behaves, these detractors do not further threaten both our morality and its comprehensive sagacity.

* * *

Turning our attention to *emotions*, where any form of Artificial Intelligence would also be severely challenged, consider the following sentence: "People are starving and homeless after severe floods," relayed by a robotic voice or a dispassionate human voice lacking complex facial expressions. This would inarguably not have the same impact, pliancy, or outcome as an emotionally charged inferential human voice; or with the situational variability seen in a primate belonging to a specific sub-group calling a mate. Consider the above statement relayed by perhaps an empathetic healthcare worker on the scene, a callous conniving businessman whose primary concern is based on minimizing insurance claims, or the sanctimonious display from a politician motivated by ego and power and a pending re-election, or a robot counting the casualties. The diverse interplay between the Logical sphere, Physical sphere, and Metaphysical in different situations between various individuals will vary significantly (who, what, where, and when) and then be markedly affected by their background and relation or interests in the issue. Further complicating this already complex interconnected and responsive network, it is simultaneously fine-tuning emotions to suit future needs. We can see how connectionism is perhaps too reductionist an approach for an enormous and perhaps even unrealistic task. It may even be dangerous in the extreme by setting stereotypical behavior based on predictable outcomes in a set normative and narrow timeframe.

Words and emotions act together as evolutionarily acquired, situationally attuned, adaptable and pliable methods for communicating and valuing objects for subjective or fallacious value in our constantly evolving enclaves—reacting to other living things and changing objects with concepts of them shifting in time. There are innumerous examples we can use to prove such interchanges, all daily happening around us. Without exception, they are purpose-driven to increase social structures, awareness, and survival, with pragmatic needs for a pliant rhizomatic cognition to continue to evolve and expand. This is not species-specific; picture a dog welcoming its owner coming home, or even intra-specific situations involving emotion directed at an object.

Inarguably, words and emotions not only enhance interconnection but also increase moral demand. It can be seen how Artificial Intelligence, lacking and singling out pliancy and emotion among other shortfalls, is still far from matching a progressive, perceptive, and pliant morality. With a perceptive evolution in its own individual genetic uniqueness, it is yet interlinked to past, present, and future in a progressive epistemology as part of a complex evolving network. However, if such new technology is wisely employed in a sufficiently evolved *moral* society, it can act as a bastion to assist in creating a better world for all of us and help us to explore the uncertainties and unknowns of an unfolding universe—and not each other.

## 3.2 Evolution of Emotions

With current understanding in both evolutionary neurophysiology and genetics, evidence of phylogenetic 'pre-emotions' even in brainless unicellular organisms, such as the swamp mold again, is well recognized. And even here, if over-inflating the potential of a connectionist model, we will struggle to accommodate the diversity of the adjustable responses needed to accommodate trans-genomic variability, unpredictable change and the adaptations needed for the continuity of transgenerational perceptive life. These little molds discordantly sense, express (agitation to light), respond, and group together, and have the means for plasticity in their DNA (and potential in a surplus DNA) to produce a variety of proteins in response to changes, in what can be at times capricious environments. Furthermore, with new understanding in neurophysiology and evolutionary genetics today, we now see progressive complexity in primitive emotive behavior escalating together with perceptive abilities as the basis of evolution. Such ancient trans-genetically attuned precursor emotions were and are all important, relevant and essential to interconnect and assimilate knowledge about an equally progressive environment—that is why they remain. Not only to increase the odds of group survival, but most importantly to aid an evolution directed toward what is now becoming *a priori* plain—advancing an interconnected perception, morality and awareness in a living cosmos. In its more complex form, they granted us the origins of the 'emotional' human brain with its complex environmental interactions and ability to recognize the interconnectivity of it all and to better respond to each other's needs in a living network.

These basic little 'pre-emotions' can clearly not be written off as mutative off-chance, disregarding events. Neither do they satisfy if seen extra-bodily or subjectively to explain the amazing persistence and order needed to result in the co-evolution of receptors like the human eye, able to trigger an array of emotions amassed in a complex brain.

We can briskly mention an often-used example here as seen in the fruit fly, *Drosophila* species. It has a very primitive brain and yet is quite able to sense a multitude of objects and respond by means of different emotions. "Flies have emotions," as neurologist and author Antonio Damasio insightfully claims in his book, *Looking for Spinoza* (2003, p42). Damasio asserts that one of these little flies, like many other lower-level organisms, "gets tired, requires sleep, has the ability to become fearless if consumed by alcohol (ethyl alcohol), and appears to be happy when exposed to sugar."

During the Cambrian period, evolutionary advancement of cup-shaped eyes much improved the means of light (photon) reception from those available to our slime mold. As an aid in determining both the direction light is coming from and the movement of nearby objects, it is now a well-recognized fact in evolutionary biology how such advancements have passaged and progressed trans-genetically. We can also now in the context of the above relate to the eye's ability in helping to formulate more complex emotions and ideas about our world as simultaneously advancing from those of single-celled organisms and fruit flies. These improvements needed only minor adjustments to the basic evolutionary building blocks masquerading in nucleotide positions of the DNA as discussed above—in a united drive. Such evolutionarily advanced eyes progressed and developed with subtle nucleotide changes stimulated by perception of changing environmental needs, resulting in connections to muscle cells—again, advancing from unicellular organisms previously dependent on movement through structural membrane changes in these pseudopods or waving cilia and so on. All such significant progresses in communication methods are made by a few small shifts in alleles in our DNA and RNA, driven by a demand to better perceive and interconnect.

Increased vision and mobility subsequently resulted in improved exposure to the environment, reciprocally with more opportunity for interaction, exploring and idea formation—these vital components simultaneously enriching and subtly altering a progressive Physical sphere. Driven to interconnect with each other and attune with changing environments, the ability to express more complex emotions and ideas serves a pragmatic Physical sphere well. With emotions as an integral part of cognition, operating within a perceptive network, emotions are the outcome and essential need of an ever-advancing interconnected perception.

More recently, and offering further support for our argument here, zoologists have also studied and recorded octopuses engaging in play behavior, solving mazes, and opening jars to obtain food. These emotions and actions are seen in descendants of mollusks with a maximum lifespan of four years! Expansive knowledge in neurobiology today about our evolutionary cognition is based on research done on such basic primary emotions in elementary organisms. There is perhaps no need to further discuss the complex issue of neuro-emotional evolution in this text, as it is well covered elsewhere. What is important to us, and without much more support needed, is how everything is indisputably more interconnected, pliable, interdependent, and *perceptive* than when set in a reductionist model of natural selection.

## 3.3 Emotions and Morality

Emotions are directly involved in the formation of morals and ethics as the following few examples may demonstrate:

- Emotionally charged clapping of hands after a satisfying performance at the symphony
- Crying or expressing sadness and empathy when witnessing a starving child
- Anger when we see someone stealing money from a blind beggar
- Inappropriate emotions, such as laughing when one sees a vulnerable young child or animal being hurt
- Close to home today, the wavering emotions triggered by the at times asinine statements made by prominent world leaders.

If any of these above (few of many) variant expressions functioning in unison with a complex environment are opportune, then emotions have a crucial role to play in ethics in revealing to us something like unavoidable moral facts.

Traditionally, according to Ekman, Davidson and Friesen (1989), the human emotions are happiness, sadness, fear, anger, surprise and disgust—the latter two of which were considered by some traditional researchers too simple to be called emotions but are now revived as only part of a much more complex definition. The neuroscientist Antonio Damasio defined emotions more objectively:

> Emotions are actions or movements, many of them public, visible to others as they occur in the face, in the voice, in specific behaviors. To be sure, some components of the emotion process are not visible to the naked eye but can be made 'visible' with current scientific probes such as hormonal assays and electrophysiological wave patterns.
>
> A. Damasio, *Looking for Spinoza* (2003, pp24–8)

Damasio's more contemporary and objective view perhaps emphasizes here the obvious error in any attempt to oversimplify and categorize emotions as traditionally set by psychologists.

Historically, the Greek philosopher Plato, in *The Republic* (380 BC), saw three basic components of the human mind: the reasoning, the desiring, and the emotive parts. Some recent work, often drawing support from the burgeoning study of the emotional brain in psychology, has suggested that while emotions typically involve both cognitive and conative (desire and impulse) states, they are distinct from both, if only in being significantly more complex. More recently and more objectively again with functional MRI studies, it is seen that one cannot separate cognition and emotion. Based on such understanding we cannot consider emotion separate or as either more complex or simplistic than the whole interconnected anatomical and physiological brain working in synchrony, pliantly interacting and responding to complex changing environments, while recruiting both.

The question for us here is not at all *whether* emotion is an evolutionary consequence with value to social structures but *where* and *how* it could function more effectively. Is it in our Physical or Logical sphere or perhaps, most likely, a Logical sphere interplay with predominant Physical sphere consequences?

Equipped with an emotional brain primed to moral behavior and with a now more sober outlook on ourselves, we need to perhaps be again reminded of Kant's view that as emotional beings our reasoning is set to be on a "risky course," *gefährliche Mittelstrasse*, halfway between wisdom and irrationality. Kant went as far as calling us "cosmically mediocre."

We can now understand how we as emotional creatures are persistently torn between ideas entering and exiting between our Physical and Logical spheres and how emotions may at times affect our rationality, with a justifiable reason—to better communicate. Although perspicacious in design, emotions can at times with relative ease be manipulated by false beliefs or fear created by dogma or by dangling financial or other forms of reward or paucity.

As an example, to appear saddened by the growing poverty gap because it is politically correct while simultaneously deep-down feeling secure and superior in your own financial comfort is specious and delusive. Or alternatively, to iniquitously suppress your anger while seeing how the corruption of a few individuals affects others and remain emotionless to protect your personal interests. This does not call for an aggressive response to others' wrongs but is a reminder of our moral obligation in employing truthful emotions to partake in an intelligent move toward active change that may improve the human condition by triggering social change. This demand calls for a responsible exchange of emotions between the Logical and Physical spheres. Emotions, in creating awareness of underlying social issues, can then help us become "cosmically less mediocre" when appropriately expressed. Thus, interconnected with moral obligations and when governed by selfless understanding, we can respond with emotionally acuity. In turn, deceptive emotive behavior is decidedly morally wrong. It can clearly be seen how ambiguous emotive behavior used by sociopolitically influenced media or prominent figures to sway outcomes may be a form of social bullying—not necessarily the intended purpose of emotions evolved to alert us to authentic dangers or social wrongs and conducive to our moral path within a network.

* * *

I hope I have now, with more than enough support and clarity in my argument so far, shown that a disregarding, insensitive, and dispassionate evolution (a greedy unprincipled DNA with no 'emotion'), not in search for pragmatic outcomes, is not viable. Neither would it suffice to explain the complexity and diversity of life, or the resilience of our higher cognitive abilities to interconnect and constantly advance innovative ideas in an evolutionary epistemology. We now understand a more ethically interactive evolutionary process aimed at creating a platform for the progression of a principled perceptive network to expand and grow, with emotions as a functional 'genetic idea,' as part of such a process. The idea of an insular, unprincipled, and insensitive evolutionary drive can now be moved into our Logical sphere of reasoning as an outdated concept, such imparity incongruous with new evidence and understanding. Perhaps it will return to the Metaphysical in time. It was such a hopeless concept, I believe, that also gave the 'adapt-or-die conniving ape' evolutionism a bad flavor and is still responsible for our interpretation of the world as an unnecessarily harsh and even necessarily immoral place, set in limitations. This outlook, wrestling to sway opinion or manipulate belief, is also the main characteristic of our economic system and in politics today, responsible for ongoing conflict and injustice. To sway emotions for personal gain by inciting fear in setting shortages whether real or not, proposing unrealistic threats, or when better suited, staging unrealistic confidence is evidently the current order of the day.

With the more intelligent and benevolent view of our evolution now, approaching a post-technocratic society, I believe avoiding the emergence of a pseudoscience is also more procurable. Global harmonious coexistence based on pragmatic solutions with fair outcomes has also never before been so within our reach. It can also shield against the unsubstantiated and senseless attacks on natural moralists, environmentalists, and other pragmatic free-thinkers and writers creating hope in progressive ideas for an emerging society, in what can be with sufficient evidence all around us only described as anxious times.

Secondly, we can avoid practicing merely a utilitarian morality driven by 'staged emotions' in a Physical sphere of reasoning while being manipulated by economic and political dogma or false beliefs. It now appears ignorant and naive to hope for any meaningful and pragmatic normative to appear from such hubbub, and all we get, inevitably, is ongoing conflict. We also have more options to emerge as a tolerant and more peaceful society if we accept the constant principled variability under which morality must operate.

Thirdly, we can only realistically reduce the complexity of what is good for the whole and *how* to define the whole, once we define our cognitive wellbeing as a unanimous, interconnected, and universal concern. While discriminatory behavior cannot be considered as a virtuous moral value from any angle, sensory wellbeing based on current revelations should now include a definition to cover all lifeforms, *inclusive* of Earth as our living and perceptive mother planet. I cannot think of any other approach than naturalism, science, and a morally adjusted sociopolitical system, working justly together, to achieve this and tackle the pressing environmental and global sociopolitical issues we all inevitably will have to confront. Neither can one avoid thinking that the purpose of advanced cognition is to *escalate understanding* of objects and how they interact under such a higher morality in their search for a better world and future—and so reveal the infinite potential and wisdom of a vast universe.

A new revolutionary life-formula built on $\infty\Delta a \approx \infty\Delta b$, where countless interactions play a role in the transmission of changing ideas, not only supports a just morality but also demands it. We have listed the evolutionary moral rules of:

- Necessity for failure
- Interconnection
- Pliancy
- Pragmatic idea-selecting, circulating reliably between different spheres of perception
- Functioning within and as part of a progressive principled perceptive drive.

Promptly replacing the reductionist concept of life seen as an antagonistic struggle to survive and reproduce, governed by fixed ideas within a set normative, is vital to our progress and survival as a species. And in our privileged advanced state of perception under new understanding, an entire living network. Facing an era of multidimensional concerns with a bombardment of constantly changing ideas at the click of a computer button, an adjustment is urgently needed if we want be prepared, both perceptively and morally, armed with sound knowledge for the complexities of a post-technocratic society. Such an interconnected brave new more consciously interconnected world, reiterates the vital need for an ethos of tolerance and respect, as is stated in a progressive, multidimensional principled cognition. Entrancingly, a progressive morality is still dangerously overshadowed by blind profits in a myopic economic model with its fixed mindset. Becoming masters at profit making may however ill prepare us for the new future we have already entered, at least under newer understanding in the biomedical sciences.

Inevitably, again we can only argue from where we stand with higher cognition as our principal evolutionary achievement and reciprocally cannot abandon naturalism without ignoring our origins. Claiming enhanced cognition as vital to our evolutionary success, we are also well insulated from being called idealistic in deducing that the purpose of evolution is to advance and nurture a progressive cognition and so strive to improve the morality that comes with it. Survival is an important and crucial issue, but much more resolute than an evolution centered on survival only, is one based on advancing morality and cognition as its core motive. The urgent action and moral demands this calls for on a global level cannot be ignored by any but those blinded by narrow rewards or the completely self-absorbed in a materialist world. Therefore, also, come the desperate and prudent calls from many healthcare workers, educationalists, and environmentalists to urgently place more emphasis and effort into setting up bias-free, universally acceptable and more humane healthcare and economic systems—freed from the current financial constraints and biases. It is pressing that these vital and valued parts of our social development should be free from dogma and manipulation by fisco-political powers or religious biases (or any greed-driven enterprises) shielded by financial power.

It has been claimed by some current-day philosophers that morality comes into existence out of the need for cooperation and so sets the terms of it (Browne and Milgram, 2009). Based on the pliancy, interconnection, and variability seen in biology today, it is impossible to consider a workable evolution where cooperation, altruism, emotion, and respect are not key ingredients, and higher cognition not the principal motive to improve understanding between *a* and *b*, and co-evolve as part of an interconnected principled cognition.

Although we cannot today agree with Kant on his stance about the morality of animals (who, *when* and where you are), full respect is due to him for his understanding of the universality of life. In the *New Elucidation*, toward the end Kant states:

> Since we find all things in the universe to be interactively connected...The same pattern of the divine understanding, which generates existence, also establishes the relations of things to each other, by conceiving their existences as correlated with each other. From this it is evident that the universal interaction of all things must be due to this...divine idea.
> (1:413.13–20)

We can now in conclusion in this section claim the following. Evolutionary cognition together with primitive pre-emotions, favored by a natural selection, progressed to form the very basis of cognitive advancement and so set the evolutionary first steps toward the development of more complex and interconnected emotions. Organisms exist in their own objective Physical sphere but are vitally interconnected and dependent on their Logical sphere and the Metaphysical for survival and the constant creation of new ideas—emotions play a vital role here. Perhaps in the extreme we can claim that evolution is both perceptive and compassionate.

* * *

A 'disregarding' survival as a primary aim of evolution, in the context of the above, inevitably must be considered against the alternatives of extinction as a casual or necessary outcome—a rather pointless topic in any sober argument. Such a disregarding emotionless drive, or an evolutionary drive *not* seen as partaking in a unanimous principled process for advancing knowledge in progressive complexity with escalating moral concerns, cannot be called evolution.

If extinction is taken as an inevitable outcome of a rogue drive, there is also no point in further discussions around moral concerns here or anywhere else, so we can exclude such a detrimental concept at once as an aimless path leading to a barbaric end. Considering extinction casually as perhaps a probable outcome of our current state is equally unhelpful. This approach is submissive and immoral because it shows lack of concern for our future wellbeing. It also shows disrespect for the enormous effort and sagacity needed to advance cognition and our epistemology to its current stage. This may also further promote disregard for our world and the welfare of others and promote immorality. Furthermore, an unperceptive senseless evolution, without hope and emotion, will *a priori* favor extinction, so we must concede that survival is an essential need, and interconnection and cognition with escalating moral concerns, the selfless aim of an evolutionary drive. With the concept of an escalating morality as an essential co-requirement for the continuation and progression of life, we have both hope and potential to advance a civilized society into a post-technocratic society with boundless potential.

We are more than merely genetically primed, but also duty bound, to *wisely* nurture our evolving knowledge and compelled to do so morally, in full regard of the perceptive network we evolve our understanding from. There is no place or time for external moralism or internal fixations, doomsday theories, or egoism here. The reason, I hope, by now is clear enough.

### 3.4 Popper and the Metaphysical Dilemma (PMD)

It has previously been noted by others (Campbell, 1974; Wuketits, 1985), and reemphasized in my argument here, that the flow direction DNA→RNA→Protein→Organism was inadequate and limited in its ability to explain the complexity and diversity of life. Campbell was possibly the initiator of the term *evolutionary epistemology* but Popper and Eccles had a more pronounced impact in the earlier application of this concept. Popper and Eccles suggested life is an 'active and interconnected phenomenon,' constantly in cognizance (perception) of its environment in an ongoing exchange of ideas with emotions and sense as invaluable compatriots. Internal and external selection and adaptive systems work in synergy. (K.R. Popper and J.C. Eccles, *The Self and Its Brain: An argument for Interactionism, 1977.*)

Since Popper's and Campbell's work, we have seen and discussed here now how more recent research in neuroscience and genomics have added the concept of a *mobile* DNA (besides replacing the order of DNA and RNA) to create a much more interactive model in support of Popper's statement and also our current proposal here. This concept now finds objective support in recent scientific research, seen in a continuously intercommunicative gene-pool able to incorporate newly formed nucleotide exchanges and result in an RNA with vast protein-producing capacity and diverse potential. Potential also escalates when understanding these interactions as operating *within continuous-*

*ly changing environments*. This rids science from its previous fixtures, now freed in a principled evolution aimed at harmonizing interconnection, rather than aggressively exploiting a world for self-directed survival in a despotic (if not meaningless) struggle to compete.

We now have RNA⇆DNA⇆Protein⇆Organism as a new basic platform in evolutionary biology in an escalating interconnectivity and perceptivity. Cognition becomes the unidirectional progressive drive of pliable interchanges in a living and perceptive network. We can also place the above bidirectional flow of organic life in a more progressive interconnection of ideas aimed at advancing perception. With more pliability, continuity, and intercommunication evident in nature, a reductionist unidirectional self-centered evolution is now firmly sealed in its coffin. With evolution seen as motivated both internally and externally by a progression in cognition and morality, it co-emerges as a rhizomatic and progressive perceptive network. We can formulate the flow here as:

> Molecules⇄RNA⇄DNA⇄Protein⇄Organism→Interconnection→(Cognition⇄morality)

It is important to note that cognition is neither uni- *nor* bidirectional, but multidimensional, and is a changing temporal activity dependent on where, when and who, all progressing in time. Cognition and perception (represented by C in this text) now appears as an infinite change and can be expressed as = $\infty\Delta C$, where RNA⇄DNA⇄Protein⇄Organism operates trans-genetically within an evolutionary formula that can be stated as:

> $\infty\Delta a \approx \infty\Delta b$ unfolding a progressive cognition, $\infty\Delta C$, in a network growing in complexity while evolving unknowns

As an interconnected constantly changing affair, the evolutionary process can then be proposed to be a product of:

$$\text{Evolution} = \sum\infty\Delta C\{\infty\Delta a\approx\infty\Delta b\}$$

With such a new outlook and understanding, it is possible to sense the potential of achieving some level of global universal neuro-eudaimonics—where renewed focus is placed on wellness seen in terms of the degree to which an individual is fully functioning as part of an *interconnected cognitive whole* instead of an isolated anatomical (somatic) structure. These two views have given rise to different research foci and a body of new knowledge emerging, aiming at a healthy and prosperous human state, in equilibrium with its surroundings and place in the universe; perhaps still in need of more objectivity, this may now *not* seem so 'alternative' or far-fetched anymore.

* * *

We can clearly sense the urgent need to stop 'fighting' for survival, competing for resources, and depriving each other of access to a better world by isolating and fixing knowledge and resources in a creative universe with infinite perceptive potential. Relying now on a perspicacious evolution to open new doors as our perception evolves in this infinite wisdom, there are no more battles to fight, only knowledge to pursue and share while we attempt to understand the principles of these interconnections in its growing complexity.

Karl Popper persistently suggested in his works that any system is an *active* system *in search for a better world.* This concept became the principle of *In Search of a Better World.* Although biological science has much progressed since the late Sir Karl with genomics and all, his insight into this interactive evolutionary world is a wonderous experience. We must understand that fewer options were available to biologists at the time, and evolution was still set in the concept of a fixed blueprint of the newly discovered Watson and Crick DNA helix. This progressed, changed, and advanced to our current concept in biology, to where we stand now with a more flexible DNA, open and perceptive to change. Much has changed now, liberated from the previously genetic blueprint, 'adapt or die' concept of before.

Popper's core proposal, that knowledge can never be absolute or perfect, led to his equally famous theory of falsification. Popper also proposed the novel idea that the growth of science is a cycle of conjectures and refutations (*The Logic of Scientific Discovery*, Routledge, 2nd edn, 2002).

Popper's falsification theory, albeit prior to the current knowledge of the plasticity and mobility of DNA in living organisms, fits our proposal where a universal cognition is constantly in search of a better world. With an evolution now $= \sum\infty\Delta C\{\infty\Delta a \approx \infty\Delta b\}$, and in a dynamic environment needing pragmatic values and outcomes for harmonious coexistence, ongoing falsification of ideas now becomes a necessity, not only to a true science, but also to a progressive evolution. Popper's thinking was further backed by Riedl (1984) and many others proposing 'a flow of cause and effect in two directions in any ecosystem where a central dogma cannot express the ultimate truth.' This was also prominently reiterated by Franz M. Wuketits (1981).

Both us and our ideas can also be interpreted as an idea that is testifiable and falsifiable, flowing in a multidimensional and pliant cognition. The metaphysical (unknown), as a source of potential ideas, serves then as a *drive* for the evolutionary process. To understand this concept a bit better, consider the following:

The ultimate truth (finding finality in knowledge) will extinguish the metaphysical and all its unknowns. With no more unknowns and existing in a Physical sphere with complete knowledge, the Metaphysical as well as a multidimensional evolving cognition will, based on simple logic, have no motive or purpose to interconnect and form new ideas. Perhaps this is why we hear at times the pretentiously erudite claim by some that life has no purpose. Devolving now, it will lead to eventual extinction of all perception. In the unimaginable scenario of having no more unknowns, PSR⇌Metaphysical⇌LSR is limited to the PSR⇆LSR. Eventually set in a theory of everything (what we so ambitiously strive for), there will be no need to change. Idea-making without any unknowns or change is obviously not possible; there is now also no need for the Logical sphere anymore. Imagining living in a Physical sphere with everything known in a 'theory of everything' is clearly impossible if not intolerable. We can see how we are completely dependent for our existence, life, its beauty, and thoughts, on change in an infinite flow of evolving unknowns in a multidimensional cognition. Unfortunately, syllogistically we also have no means to leave our Physical sphere and start reasoning in the Metaphysical without ceasing to exist in our Physical sphere. It would be hard to look for an unknown black cat in a dark room with no means of perception, and who would care anyway?

This absolute need for a principled interdependency, interchange of ideas, impermanence of knowledge, change, unknowns, and falsifiability operating between the Metaphysical, Physical, and Logical spheres, we can perhaps refer to as the **Physical sphere/Metaphysical Dilemma**.

* * *

As we can gather from the above, without cognition, morals (all-regarding conduct), and emotions, all interchanging between these spheres of perception, it is not possible for evolution to function or for perception and ultimately the human brain to evolve. Even if we could exist in such isolation in a metaphysical world, we need perception, operating in a Physical sphere with input from a Logical sphere with all its discrepancies and uncertainties operating in unison to continuously evolve the Physical sphere. If not, we cease to exist as perceptive and interactive beings, aware and alive in an objective world.

Deductively, from this there cannot be a senseless world. We are therefore entrapped in a vital Physical/Metaphysical Dilemma where we cannot evolve a trustworthy Physical sphere without the enthrallment of the unknowns of a Metaphysical. We exist as an evolutionary cognition interconnected and magnetized by an infinite, equally progressive, Metaphysical.

Seen from this unique perspective, we are bound to follow a progressive morality to progress on all levels, for trustworthy interchanges between the Physical, Logical, and Metaphysical—driven to increase our understanding of simultaneously evolving unknowns. Our infinite evolutionary cognition can now be expanded on and be expressed as:

Moral Evolution or Ev(mo) =

$$\sum\infty\Delta C\{\infty\Delta a(\text{Metaphysical}\leftrightharpoons \text{LSR}\leftrightharpoons \text{PSR})6 \approx \infty\Delta b(\text{Metaphysical}\leftrightharpoons \text{LSR}\leftrightharpoons \text{PSR})6\}$$

*C=cognition; 6=the aforementioned orders of arranging PSR, LSR, and M

Based on this formula we may conclude there are no ultimate truths or complete knowledge, only ephemeral workable interactions. And should complete knowledge ever be concentrated it will most likely result in the end of perception—perhaps the contraction and void before another Big Bang and the reason egocentricism and arrogance is literally senseless.

Evolution creates its rules, both *a posteriori* and *a priori*, in a multidimensional cognition. It progresses rhizomatically with increased perception. However disheartening this may seem for those in search of comfort in familiarity and ultimate truths and theories, it is important to realize that unknowns and uncertainty are the key to our ongoing search for knowledge and understanding of the world and universe around us. Our Physical sphere remains as 'pliable' security, yet vitally dependent on both the Logical sphere and Metaphysical to pragmatically progress. Our *morality* (principled interaction) *is the only security* we can utilize to further evolve and get closer to temporary 'near truths,' which is the best we can do and are designated to do. Egocentrism and material fixations may one day become the embarrassing hallmarks of a flawed reductionist era.

The human brain is certainly not some sideshow of a *mutative* 'freak' event governed by self-serving rules to adapt or die in a disfranchised natural selection. We furthermore also cannot have fixed rules or a normative without dogma, and clearly now there are no complete or non-falsifiable rules or a fixed normative in evolution. We are driven by an infinite moral particularism with interchanges in a multidimensional cognition in search of improved knowledge and morality to create an ever-elusive better world.

*Evolution in its highest design is now rather the ethical and pragmatic demands set by a multidimensional preceptive drive*. It employs a malleable natural selection following a moral code of interchanges to draw objective realities *a posteriori* and *a priori* into a pragmatic and infinite idea-forming process. Not the other way around. This process is constantly advancing interconnections and complexity, not only within our world but also as part of an expanding (cognitive) universe. We are inescapably caught in this progressive cognition, no matter how small or insignificant we may think we are in the context of things. Religion, natural selection, and Grand Design then seem to be different ideas and reductionist concepts all struggling with the same issue—some striving for more objectivity while others surrender ideas to the equivocality of the Metaphysical, yet all as part of an ongoing idea-making process and evolving epistemology—as long as they can adapt and progress.

There are many, mostly creationists, who still consider science, evolution, and medicine as audacious, trying to overshadow the impact of religion and to segregate society from a God secured in the Metaphysical. I think in view of my proposal here, working in interconnected spheres of reasoning, looking for rationality and morality, that science stands out as an ethical duty. It needs all of us (certainly not only scientists) to harmoniously interact, be humble, unbiased, pragmatic, and to progress as a united concern. If this science stays ethical, truthful, and driven by a respectful search for ephemeral values and truths to create a better world for *all* of us, it will be partisan to a universal cognition.

We must reemphasize that DNA did not, by a rather crude and selfish process, with a bit of luck thrown in, favor genomes to eventually set up cognition and emotions and bring organisms together that can communicate. Besides being driven by sagacity in mobile nucleotide combinations that had to be perceptive to their surroundings to survive, it also had to be sensitive to its moral obligations (in full respect of others and the environment) or face permanent extinction. The archaic Darwinian view of the human intellect, objectively interpreted as a mere off-chance event tested against the harshness of an environment and critically valued merely for self-serving purposes to survive and reproduce, may have been elementary and superficially attractive enough in a reductionist past. However, we should now acknowledge an ethically evolving knowledge-gaining process as a moral responsibility to reduce further suffering in our search for a better world. In facing the infinite potential of a simultaneous progressive Metaphysical, this is clearly an innovative and more enlightened choice. After all, it is through our cognitive abilities and emotions that we will suffer most intensely if we do not acknowledge our evolutionary cognition and its vulnerability and potential, and not through material lack in a universe of abundance.

## 3.5 A Principled Cognitive Evolution

*Nothing in biology makes sense except in the light of evolution.*
Theodosius Dobzhanski (1973)

A 'new' evolution is now more than well enough armed with objective evidence to confront the inadequacies of creationism, grand design, a reductionist natural selection, or other false beliefs. It now also comes with renewed moral requirements to adjust some archaic beliefs still diverting our morality.

Evolutionary biologists postulate the mechanisms that can lead to changes in allele frequencies (genomic-level method of passing on new genetic ideas) to be natural selection, genetic drift, genetic hitchhiking, mutation, and gene flow. These five methods currently aim at explaining how genes (using the subunits of alleles and genomes) manifest themselves to a constantly changing environment or ecosystem, creating the opportunity for new 'genetic ideas' to emerge.

**Natural selection** is inarguably self-evident, easy to accept, and difficult to reject. As part of a larger truth it narrowly states that traits that are favorably matched to a certain ecosystem will continue to be selected and continue into new generations, and those that are not fade—in this simplicity also lies its security. We now know that they may also be contained in the genome for future use or altered by the epigenome in the living organism.

**Genetic hitchhiking** is, to us here, the exciting ability of an allele (a variant mostly due to a mutation of a specific gene) to change frequency not because it itself is influenced by natural selection, but because it is *near* another gene on the same DNA that is undergoing an adjustment to its surroundings. It is a form of neighboring gene intercommunication and interchange of ideas.

**Mutations** or changes in the DNA sequence of a genome are also now, through better technology and with new understanding, increasingly seen as *not* random events but stimulated by interactions with the ecosystem of a genome, an outcome we could now almost expect to be the case after revelations here.

**Gene flow** in turn is a term used for the genetic exchanges taking place between populations and species. In extreme cases and as an example here, bacteria can cause genetic flow in having the uncanny ability to exchange genetic material between each other without any offspring; hence the relative ease of antibiotic resistance seen in veterinary and medical practice. Viruses also have the ability to introduce DNA directly into other cells, mostly causing diseases but also more recently showing significant potential to transport beneficial alleles to damaged cells. There are still more examples of this, also referred to as **horizontal gene transmission**, occurring in nature. Talk about interconnection!

**Genetic drift** is another more recently studied method used to explain the diversification of genetic ideas. Defined in simple terms, this is when the occurrence of variant forms of a gene, called alleles as mentioned above, increases and decreases by 'chance' over time. These variations in the presence of alleles are measured as changes in allele frequencies. This opinion is under strong attack and changing as we write, since even in a fixed population and ecosystem, newer understanding is now revealing a more mobile DNA than previously thought. Nothing happens by chance but rather due to *interactions* and communication on many levels, while alleles are exchanging genetic matter or their products (proteins)—perceptive to changes and open to formulating innovative ideas. It appears that Riedl and other likeminded thinkers at the time, perhaps more philosophically, were right about this interconnected, living web.

What clearly stands out in all the above is that genetic activity is based entirely on constant interaction, change, and intercommunication on numerous levels—this communication stretching transgenerationally.

Our interpretations of evolution can subsequently be inferred in the following ways, where Ev=the evolutionary process, and C=cognition, LSR and PSR have been explained, and the rest is well recognized as standard mathematical symbols:

A. **The historic reductionist view**. Here the human brain and cognition are seen as favored by an obtuse genetic code operating selfishly under the mechanical laws of a rogue natural selection. Fundamentally valuing survival and reproductive fitness, while mechanically matching a disregarding evolution and its genetic matter to a harsh environment, the human intellect is considered as an adventitious (emotionless) event. With human cognition seen as an adroit tool for exploitation of a harsh world in a disregarding evolution, we can see how this now-disjunct concept can promote egoism, material fixations, monism, and extinction theories in a selfish battle for solitary survival. Such a reductionist approach sees cognition primarily as a disengaged 'naturally' selected benefit for despotic benefactors. With *a*, and cognition as coincidental and an isolated consequence to manipulate and confront an antagonistic *b*, this could be formulated as: Evolution=$\Delta a(C) \Leftarrow \Delta b$. This is now convincingly seen as inadequate and even diabolical.

B. As suggested here, a **principled evolution** and its genetic building blocks are considered in acknowledgment of natural selection, gene flow, genetic drift, mutations, and genetic hitchhiking as coterminous in advancing complexity in a perceptive network. We now postulate a more universal interchange of constantly changing interconnected and interactive ideas in a more comprehensive formula. Where a principled evolution, Ev(mo), and cognition/perception, C, is seen as:

$$\sum\infty\Delta C\{\infty\Delta a(\text{Metaphysical}\leftrightharpoons\text{LSR}\leftrightharpoons\text{PSR}) \approx \infty\Delta b(\text{Metaphysical}\leftrightharpoons\text{LSR}\leftrightharpoons\text{PSR})\}$$

C. **'Grand Design' and belief systems**, with no evidence of their core divinations. These remain a search for security in a solitary, if not ominous, Metaphysical. Inevitably then lacking enough empirical support or any method to confirm or be falsifiable in a Physical sphere, it will remain a vacuous topic in the Logical sphere.

D. **Options we have not considered because they still escape our current perception**. Remember, the more our understanding grows, the more the Metaphysical evolves and the more our understanding will progress, and concepts appear and disappear in our evolving epistemology.

There is a remaining and worrying trend among many contemporary thinkers, namely to still be entrapped in a reductionist mechanist version of natural selection (choice A). This is most likely so because, besides the ongoing impact of tradition and culturism in both general society and academia, it avoids shaking existing foundations, where many current theories are still aimed at conclusions and finality. And then there are those with still one or both feet in choice C, struggling to accept even the *initial* concept of evolution, trying to fill inadequacies with metaphysical beliefs.

The reductionist approach could also, when required, completely steer clear of the metaphysical, now seen as an unforgivable mistake, as we have pointed out here. With most of the new science still to be assimilated, it will also need time for adjustments to be made in generalized thinking and education. This, however, is bound to change sooner rather than later, set now in an era with an interconnected world opening many new avenues of communication and potential. There is also an urgent need for social change driven by a global realization of pressing environmental issues. This will further speed up adjustments and place pressure and demand on decision-makers. Such an adjuration is begging for appropriately adjusted and *ethical* economic and healthcare systems to follow.

If choice A becomes established in our Physical sphere as an acceptable belief system (belief systems will be referred to again in chapter 5) we have, as stated, numerous concerns to address. I hope the presentation here is, besides its logic with scientific backing, attractive enough to the rational mind as an idea to aid in the creation of a better and more ethical, sober, and in the end benevolent world. Furthermore, a reductionist natural selection leaves us all vulnerable to the ineptitude of agnosticism and exposed to a rather meaningless existence in a lagging epistemology begging with the question of surviving and adapting for what? Alternatively, it leaves clinging to the emptiness of the metaphysical. The equivocality of defining 'fit' in today's complex world of genomics and technology has also become incredibly composite. It leaves us with a sense of seeing the futility of any ethic that is manipulated to suit an 'elite' or swayed majority overpowered by an inept single-minded controlling body.

We can assume that moral living and ethical behavior appears socially attractive to most rational minds. It becomes, however, no more than a tenet, open to sway in a fight for a rather rudimentary form of survival when set in opposing belief systems and greed-driven power struggles. In this opposition and lack of universality, inciting conflict, also lies a lack of trust and moral void.

Placing a 'ghost' behind the mechanics of option A, perhaps in an attempt to address the neglect of the metaphysical and any moral issues, will simply return us to unproven belief systems underpinned by dogma, and again merely result in manipulating the metaphysical for swaying support and maintaining control—and should things go wrong, conveniently redirect responsibility to another preoccupation.

Option B, as supported here, in turn leaves us with the well-defined, simultaneous open and universal goal, of advancing a progressive perception. Focusing on a global cognitive wellbeing and with infinite scope for evolving a pliant, better adjusted and more pragmatic epistemology, we also get closer to a progressive morality in our combined destiny. Such mindfulness will also offer more hope to prevent extinction of the perceptive organic matter that got us here.

* * *

We should finally in this section consider the following:

Under a survivalist adaptive dependent genetic code left to chance, how do we relate to an epistemology having evolved the ability to alter our destiny? We have a progressive science that gave us the ability to place a man on the moon, transplant human organs, and, with nuclear energy and genetic modification, either mass-eradicate or alter and improve life. With such a science, we can fight viral epidemics, and also predict with some accuracy the timing of the Big Bang, a threat from a supernova explosion, and the impact of climate change or a meteorite strike. Are we still naive and complacent enough to surrender to a singular set system, like our current economic hierarchy still linked to corruption, brutality, anarchy, and depravity, and ignore the large-scale benefits of all our advances? *Or*, is it not in this ability of ours to advance our morality and perception enough to *overcome* the narrow confines of a reductionist evolution and set beliefs that our duty and cognitive future lies?

So, at least for us as perceptive beings, natural selection is merely an end to a much larger means. Based on this simple tautology we can then conclude (and perhaps have no other choice) that the purpose of natural selection is enhancement of cognition to interconnect and grow our morality and understanding as a network, and therefore it is realistic to harbor concept B in our Physical sphere. This is also clearly why we have remained responsible enough so far to control nuclear war, show concern about global warming, experience sadness when seeing suffering in other sentient beings, attempt to steer clear from uncontrolled cognitive and genetic manipulation, and try to address growing poverty and suppression. Perhaps this is also why we will always remain superior to Artificial Intelligence and disapprove of greed, war, poverty, inequality and immorality, and continue evolving into the universe as perceptive moral beings in what makes us human.

We can again revise:

- If progressive cognition and understanding turns out to be the primary goal of evolution, this comes with enormous responsibility to nurture, respect, and advance healthy interconnected perceptive and cognitive abilities on all levels—expanding wellness universally in One Health.
- A detached survivalist-based naturalism is a vestigial view that creates ample scope for disregard and neglect of lower links in our phylogenetic tree, to connive, manipulate, corrupt, make war and ultimately destroy everything under egocentric survivalist strategies—anticipating inevitable extinction and moral demise as our destiny.

It should be clear from this presentation that it is more congruent with our destiny and design to function under a perceptive evolution with moral responsibilities under a progressive universal ethic. We should also be better placed to interact with a changing metaphysical in our

Physical sphere while drawing trustworthy ideas from our insecure Logical sphere. Such respect (see next section) for the perceptive interaction and interconnectivity of our existence sets a platform for advancing a society growing in both its understanding and morality. In a paradox, such coalescence will also grow more complexity and diversity, contrary to previous claims by reductionist evolutionists who state that segregation or conflict is a requirement to stimulate diversity and that we have reached a genetic dead-end.

Under choice A we now experience a troubled, distrusting, isolated, and immoral society, emerging with no purpose but to out-survive a selfish and cruel unprincipled world; while others take refuge in conflicting metaphysical belief systems. I, for one, value and will continue to grow the understanding, hope, and responsibility it takes under a principled interconnected and perceptive evolution, and the promise it gives for harmonious coexistence.

As a life-formula for me personally, as a practicing veterinarian, it has also asserted more purpose in healing malaise in creatures that cannot verbally relay their suffering but nevertheless suffer in similar ways to human beings. Seeing them as interconnected to both a perceptive network and environment makes the task more complete, with a healthcare slowly getting renowned for focusing more on profits than the empathy it evolved from.

## 3.6 Respect

This brings us to a brief discussion on respect, a key part of any attempt at moral coexistence on a relatively small planet in a vast universe. I define it in opening here as: *acknowledgment of the enormous responsibility and consideration for the variability and hierarchy of ideas in a selfless interconnected search for knowledge as functional parts of a progressive evolutionary cognition in search of a better world.*

Thankfully, we live in a world where most people still give up a train or bus seat to the elderly or disabled. Sadly, there is also a growing lack of respect, not only for the elderly but also self-respect, and the question then begs why? This may be interpreted as a form of rebellion against the growing greed seen in an anxious society. As an example, dysfunctional families, workplace injustice, corruption, and a trickling down effect of egocentric behavior seen in influential individuals and powerful authorities—all this may trigger rebellion. Respect should be seen as a necessary evolutionary recruit and a social need. Respect can be defined in many ways, and whatever concept we may have of respect, it is undeniably of major importance in life and evolution. We may have different concepts or valuation systems of respect applied to persons, to non-persons, nature, or self, all inevitably functioning within a complex perceptive network. So, respect can be seen in the first instance as obedience and subservience to a principled interconnected network. Some philosophers and social scientists on the topic of respect have asked (duty-bound to do so) probing questions when it comes to respect. In brief, we can list a few of these queries here:

1. How should we approach and understand respect?
2. What objects or sorts of things can demand respect, and which do not?
3. Do objects demanding respect have any common features that can be described?
4. What ways of acting and forbearing to act express or constitute or are regulated by respect?
5. Are there various levels of respect?
6. Are there moral requirements to apply respect?
7. Why is respect a moral requirement?
8. What are the implications of applying or not applying respect to morally and sociopolitically complicated areas like abortion, pornography, sexism, affirmative action, punishment, and so on?

To find answers to some of the above complexities in the context of my argument here, I suggest we expand our approach to respect (a complex issue) to include the following:

- Acting in our three spheres of perception as discussed, and in recognition or appraisal of persons or non-persons or self for what they are in relation to oneself and one's own spheres of reasoning as part of a cognitive network—we show respect.
- Overriding any ego or emotion from swaying dissimilarities, and simultaneously not surrendering or reacting emotionally to insecure belief systems or false assumptions.
- Subsequently, conceding to an open and interconnected circulation of *reason* as a universal concern—exchanging ideas between the Logical, Physical, and Metaphysical spheres of our perception.

As an example, and a current-day concern, we may take a corporate tycoon replacing his workforce with expensive modern technology to boost profits and gain tax responsibilities. The resulting large-scale unemployment creates much suffering and disturbance to the social order while one individual benefits. Respect is obviously not earned by such an indelicate 'entrepreneur' boosting his own wealth at the cost of others. Should such a tycoon instead, in respect of his employees, improve work conditions and fund retraining of the workers and focus on how he can sustain himself and his employees, perhaps as robotic maintenance-workers, respect is well due. Respect then flows multidirectionally into future opportunities and brings us closer to selflessness. There is nothing reductionist about respect, and in this, also its potential. Pseudo-respect, such as when respecting the boss because you may get fired even though he is a rogue, nasty and immoral person, should not be confused with respect. With fear being extremely reductionist, it disarms respect from its moral value. We shall discuss fear in the next chapter.

We may also show respect for inanimate objects such as a high mountain and the dangers it poses to climbers, turning into some form of fear and disgust when a friend dies in an attempt to conquer it. We consider it respectful and tolerant to wait for an elderly person crossing the street, but feel disgust when the same person is seen urinating against the wall across the road. We feel respect and concern for a woman unavoidably going into labor on a park bench, disgust when we find her squatting to urinate on it, and with casual displeasure take notice of a dog urinating against the same park bench, casually reminding ourselves to not sit on the same bench next time.

We can see that respect is situational, depending on who, where and what you are and again when, all with some social value. Furthermore, respect will also vary between assignees of respect. An example is the display of male or female nudity, which people in some cultures would find disrespectful to both their religion and person, while others in turn may consider it normal or tolerable.

In searching for universality in respect, this is most likely impossible unless seated in an ethos of non-discriminatory, non-judgmental, harmonious coexistence. This ideal in turn can only be approximated when existence is perceived as an *interlinked united concern* under a universal ethic, in acceptance of differences. Evolution as an interconnected multidimensional principled perceptive drive again, here, evolves respect as it advances emotions, morality, and cognition. We can also see respect as a form of tolerance. To avoid pseudo-respect, it needs to be applied free of tenet or manipulation in a trustworthy exchange between *all* three spheres of reasoning with a universal template.

If we approach our Physical sphere selflessly and aware here of our innate need to reduce suffering and support each other's welfare, the issue of respect becomes less facile. It does not matter so much who, where, when, or what it is applied to. We now have a goal-directed universal valuation system based on equality, detached from mere self-interest. Such an approach not only sets the basis of respect, but as such, also kindles moral conduct and tolerance and helps progress our understanding of how to function in a complex society on a universal level—acknowledging even its weakest links. Such a society is also better prepared to face change, adversity, and future challenges in a trustworthy and expansive cognition.

# 4

# Receptive Wealth Creation in a Perceptive Evolution

Perhaps the biggest insult to a perceptive evolution and its morality may be to capitalize on knowledge principally motivated by profits. Then, subsequently from an elevated position gained through such knowledge, continue to manipulate this knowledge and further racketeer to control and regulate the distribution of such tainted proceeds. In contrast, **receptive wealth creation**, which we should more realistically aim for, is seen separate from this as it emanates from a progressive consanguinity focused on equality and sharing of knowledge and resources—so that everyone can benefit and prosper. Arnold R. Eiser in *The Ethos of Medicine in Postmodern America* (2014, p22) recapitulated: "It should not be a surprise that pharmaceutical companies, most of which are publicly traded, prioritize profit maximization over scientific objectivity." Perhaps this *is* not much of a surprise to many, but hopefully it is a significant moral concern to those who are striving to distribute reliable knowledge, and conduct research to benefit us all justly in a progressive consanguinity, such as is demanded by an ethical healthcare system.

**Insensitive wealth creation**, as in focused profit-maximization controlled by an elect body, fuels methods to outmaneuver or control others for what becomes then self-centered short-vision benefits, with status, power-play, or protectionist ideologies kicking in. Emulously, it becomes insensitive to both knowledge and the welfare of others as part of an unreceptive wealth. Operating under such a disheartening ethos the best we can hope for is a higher ethic and morality to establish itself in those who gain control of this accumulation and to distribute it justly and responsibly. Unfortunately, we know this is seldom the case, syllogistically and as proven by history. Generally, the diabolical drive to accumulate and isolate copious amounts of possessions often goes hand in hand with the urge to have power and control over others and influence their views and destiny to suit personalized views and interests. It is also linked with a disregard for the environment.

Insensitive wealth creation can also stem from fear of suppression or adversity and then appeal with the promise of creating refuge from such tribulation. Inevitably, the genesis of such insensitive wealth creation, either way, comes with competitiveness, protectionism, mistrust, and defense strategies. Furthermore, once harbored in the comfort of its excesses, those who possess such unreceptive wealth are mostly inclined to become isolated from the welfare of others and their disquietude. In growing disparity, they are unable to see the world through the eyes of those less privileged. Now focused on their own self-proclaimed normative, they are in the advantaged position of distributing resources and practicing philanthropy based on personal world-views or beliefs. As a result, unreceptive wealth creates a breeding-ground for ongoing antagonism, distrust, and conflict. Misdirected control of knowledge can have the same perpetuation.

We would be naive to deny that corporate funding is not key to driving most research and education in recent years. Furthermore, with the ability of students (besides the obvious entry requirements) to afford 'acclaimed' knowledge or 'elite' education, and with qualifications now driven by 'market' trends, the power and influence of corporate business here has escalated to record and diverse levels. My concerns regarding this issue emerged mainly from healthcare but inevitably we cannot avoid how it is interconnected with sociopolitical structures. We can unfortunately assume, with growing support, that our knowledge and healthcare systems are heavily affected and swayed by what pays and those who sway this financial support. Those in control of funding and influence over our main economic structures subsequently have more power to express their own customized world-views and influence the flow of knowledge and how we should research this, with these outcomes affecting all of us and our healthcare.

Under such a system it would be demanded (and hoped) that those in power and control would operate, not only on higher moral ground, but under a universal ethic and apply superior 'wisdom.' Evidently such a Nietzschean 'superman' has not eventuated, and setting a moral hierarchy based on wealth is unrealistic, arguably even immoral. Deductively then, this climate will not progress morality in synchrony with a complex interconnected society and its evolving demands, nor will it evolve a well-founded epistemology with pragmatic value in a search for a better world. Applying different morals to diverse groups (a form of moral apartheid), having an elect few direct our morality and knowledge, is also not helpful as most of us may know by now. Based on such an insulated perspective of a normative, this will also not be conducive to a perspicuous evolution evolving a sound epistemology to match the progressive needs of a complex society in an expansive network.

No morality can, to remain true to its definition, comfortably accommodate any attempt to control, falsify, rig, or sway knowledge to suit personal views or the cravings of an elect group. The discomfiture of a manipulative morality and knowledge can on a larger scale masquerade as, and be driven by, extremist groups, religious sects, political parties, corporations, or an entire economic system, focused on isolating resources. In its extreme and worst form, it can be seen when employing violence or military means to then further enforce, suppress, defend, or sway knowledge to serve its own set views and beliefs. Be it hiding behind metaphysical beliefs, nationalism, or the power to control the supply of essential resources, such a system unassailably causes continuous suffering and ignores and misdirects an evolution based on universals and harmony to stimulate diversity and evolve knowledge openly, as a network.

Such narrowly focused self-interest can be further categorized as:

**Enlightened self-interest**, which socio-anthropologists define as a situation where some individuals act selflessly in the group only in anticipation of some return or reward in the group-order. Provoking nationalism fits this definition.

**Unenlightened self-interest** in turn refers to a situation where all

> persons act according to their own myopic selfishness (unfortunately, this sounds a bit like the world we live in currently). The group then suffers loss as a result of conflict, decreased efficiency and productivity because of lack of cooperation, fraudulent behavior, and the increased expense to each individual for the protection of their own interests. This directly increases the need for legal services, policing, and military bodies as we see today.

If a typical individual in such a group is selected at random, it is also not likely that this person will profit from such a group ethic or have much interest in its morality, other than for personal gain and protection. Most corporations today exploit a form of 'enlightened' self-interest, promising a reward in return for certain actions or services by individuals in the group, such an amalgamation principally driven by financial gain. We can easily see how this can fuel corruption both internally and externally. These individuals, should they not suit needs, have also recently become much easier to replace with a growing dependency on modern technology and with many job-seekers to replace them. Threatening to increase unemployment with growing distrust in the economic system, in an already troubled world economy, clearly we are overdue for an overhaul of our outdated and maladjusted economic system. Our evolutionary roots now echo this need. Changing an archaic system rather than defending it clearly seems more sensible than struggling to create confidence in unbendable injustice.

We should clearly be concerned when, working as 'enlightened' self-interested individuals in a society controlled by unenlightened self-interest (corporate), we systematically are now being replaced by an advancing technology that appears more compliant to corporate needs. Fraud becomes rife in such a myopic system, narrowly focused on profits and self-centered aims. Crime and corruption inflicting unnecessary suffering, loss of human life, and environmental damage increase proportionally in such a society. We see this in all forms with many worrying examples increasingly popping up. In recent years two major car companies cut production costs, resulting in damage to both human life and the environment with seemingly little concern, except financial and legal implications. It is unsettling to see the ease with which we now absorb news about corporations being involved in corruption or banking groups in fixing rates and so on. In perhaps an extreme case, we read about a practicing oncologist diagnosing non-existent cancers in his patients, with his only motive to make more money from selling harmful and expensive chemotherapeutic agents to furnish a lavish lifestyle. In such a distrusting society both morality and knowledge are now overshadowed by short-term profits. Novelty purchases, brand names and promotions beguile the innocent as a handy method to assist in such senseless exploitation of resources.

The question then also begs, how can we develop any pragmatic ethic under such conditions without intense policing systems or increased isolationism? Such segregationist strategies and their policing authorities in turn, also open to financial manipulation or ascendancy. Fueling an economic system driven by power and false beliefs, instead of lucidity in a well-formulated and reliable interactive Physical sphere of reasoning, has perhaps become the biggest threat to our ongoing civilization.

* * *

We can define *knowledge* as adaptable and *progressing from a perspicuous evolution, increasing our understanding of objects and their interconnections and interdependency, functional in a Physical sphere of reasoning and existence.* In a constant interchange with the metaphysical and a Logical sphere, it progressively changes and challenges ideas and understanding of matter as an evolving network. Our current complex and conniving economic system does not fit such lucidity. A sound knowledge-base consists of ideas and theories constantly tested against outcomes and remains as pragmatic and useful, only while they remain harmonized and of service within the network. The more force and control needed to secure any idea, system or theory, the less likely it is to be a durable idea with universal value.

And when it falls behind in its primary duty of keeping up with the needs of evolving a trustworthy network of progressive knowledge, it fails on *all* levels. The current economic system is increasingly matching the definition of such an unbudging failed system. Acting as a battle-ground for control of resources and power, it depends heavily on expensive lawyers and complex contracts to secure profit-focused activities, staged in fluctuating markets. The main motive of being productive in serving such a community has now become lost in the war-games of businesses operating as enlightened self-interests under unenlightened self-interests. Many other similar systems, not in the scope of this book, continue to cause strife and loss of life.

If we express a sound flow of ideas and knowledge mathematically, where flux is defined as the rate of flow of ideas per unit area, which has the dimensions of [quantity]·[time]·[area], we find we exist in an era where there is an explosive growth in both quantity and area of knowledge application. With such a flux of knowledge there is now not only ample scope for many more pragmatic contemporary ideas to emerge but unfortunately also more opportunity to manipulate and misconstrue contemporary ideas if given enough power. Time is also now reduced in this simple equation to keep up with this increased flow to balance the formula—perhaps why most of us today feel the day is too short.

In a hurried and stressed world, knowledge based on misinformation now disturbingly favors those backed by money and power, with great efforts to sway and protect information for personal benefit; pharmaceutical companies again stand out here. Instead of logic operating in a progressive ethic so vital to healthcare in an era of genomics, pharma has now become a significant concern. Well imbedded in chemical sales, they find ample support in an outdated economic model to gain more control of an emerging market in genomics, backed by financial muscle in profits already skimmed off the strained healthcare system. The uncertain impact of these chemicals on future generations has in recent years also become a growing concern among some scientists. We can subsequently expect a misdirected pseudoscience to prosper under the influence of profit-hungry corporations, benefiting from hurried decisions and even more confused consumers.

A still much-overlooked aspect of such mendacious activity is the environmental impact and unnecessary use of taxpayers' money for legal services to protect such audacious activity, constantly trying to downplay consumer concerns. Overshadowed by inane wars over resources, with powerful corporations gaining control over easily swayed egocentric and profiteering politicians, noxious activities by large companies are easily covered up. Enveloped in such greed we can also expect growing insecurity and distrust, with a faltering morality set in undependable knowledge. We can as a result expect our legal systems to become less reliable, more complex and manipulatable by financial structures.

Matters can get even more equivocal and contrived in this insult on science when we reflect on the following concepts raised by Popper and Kant. Popper, highlighting the need for interconnectivity, suggested that the objectivity in scientific statements lies in the fact that they can be inter-subjectively tested (Karl Popper, *Logic of Scientific Discovery*). Kant, well before Popper, also noted that the objectivity of scientific statements is closely connected to the formulation of theories from postulations based on hypotheses and *universal* statements. Popper famously concluded from this that there can be no ultimate statements in science, since there will always be new testable statements arising from the existing empirically based statements—congruent with our argument here. We can perhaps sense how profits and power can become reductionist and can easily sway logic to hamper a frank science and cause it to then, at its very best, become equivocal under a dangerous 'forced unfalsifiability.' We can only imagine how impossible the prospect of any progress or life would be, should we base it on equivocal, unfalsifiable set dogma. Clearly from this, and if we are truly committed to the benefits of interdisciplinarity, this can only eventuate if the economic system steps back or dramatically adjusts its current ethos and grip on knowledge advancement.

* * *

We should now ask, in a perceptive evolution progressing in time, dependent on moral conduct and pragmatism, while continuously adjusting to change within change, $\Delta C(\infty\Delta a \approx \infty\Delta b)$, and with ideas required to be falsifiable to progress: How can we *afford* to divert our time and attention by setting and focusing on short-term financial concerns as the principal target?

Likewise, in an objective science we also simply cannot accept a statement based on an idea simply being true because we were told so, and then place it in our Physical sphere. We need *unmanipulated* empirical evidence open to change, testability, and falsifiability—with all three spheres freely circulating *dependable* ideas. Simultaneously, we add to the dilemma if we acknowledge that many brilliant and valuable scientific and other statements are initially hazy, vague, and merely speculative concepts in our Logical sphere.

We also cannot ignore that all our hope, imagination, and creativity, what makes us human, lies in this Logical sphere of our reasoning in an interchange with the Metaphysical. So, we must at least attempt to be as vigilant and concerned as the late Sir Karl Popper here, that a science carrying such a high and fragile responsibility remain continuously free to circulate reliable knowledge between all spheres of perception. This should not be complicated or restricted by manipulative self-interest or 'what pays' in these delicate areas in complex times. With time now running short in this flux of knowledge and its increasing volume, we are urgently called to our duty here to elevate our morality in an attempt to match that of our evolutionary construct in search of a better world, and to avoid at all costs growing a pseudoscience.

* * *

Recently, Arnold R. Eiser, in *The Ethos of Medicine in Postmodern America* (2014), and a growing list of subsequent peer-reviewed papers by other authors, emphasized concerns about pharma afflicting our healthcare systems. Of major concern is when this also starts infiltrating the quality, honesty, and fairness of our healthcare systems, where we are all at our most vulnerable. In a profit-driven pharma there is an alarming trend to manipulate the truth and healthcare outcomes to suit economic interests. In a current setting where 75 percent of medical research journals and publications are funded by pharmaceutical companies, this data is indeed troubling when trying to establish an ethical healthcare system. The 'driven by profit' motive of these companies is quite clear. If we add to this the fact that perhaps as much as 69 percent (Eiser, 2014) of experts in advisory capacity have financial stakes in the outcome of such research, we can see the urgent need and complexity in introducing a new ethos in healthcare—starting with pharmaceutical companies as structured by the current economic system.

If the above sounds alarming so it should, but even more troubling is the complacency and lack of concern seen in some practicing health-care professionals, now also dragged into sales performance assess-ment schemes as corporate employees rush to their early retirements. Academia and politicians are also heavily affected by this issue of fi-nancial sway. It is in a sober idea-making process developing a candid science for the benefit of all, especially in healthcare, that our moral and ethical duty lies and our cognitive wellbeing and future. Any apa-thy seen about this issue can then only be interpreted as disregard for moral and ethical behavior, even fraud. The impact of such a crime will sadly be carried for generations to come both morally and on the genome, regardless of amazing scientific discoveries to follow, should we continue to turn our backs on this still much brushed-aside issue. We can clearly see that the essential foundations of our existence are extremely unstable.

We should (open to attack) single out our archaic economic system as the main culprit, where it has become a growing myth that trickle-down economies will improve our destiny. Designed and entirely focused on profits, there is no sphere for self-examination to see if perhaps there may be better or other options. In an out-of-control outdated economic model it is increasingly difficult to create an ethos that can meet the requirements demanded by an ethical science and healthcare with trustworthy exchanges between all spheres of reason. It is a mistaken and unproven belief that trickle-down economies will create our destiny or a fair society; *society will create its destiny based on an ethically connected network of progressive and reliable knowledge evolving a trustworthy epistemology.*

Our current outdated economic system can be seen as a constantly manipulated Logical sphere activity in our proposed model here. Well removed from the Physical sphere and its scientific objectivity, and reluctant to reinvent itself, it inevitably will have to be replaced by a more humane and pragmatic system to keep up with our progressive epistemology and morality. Undeniably we must accept that not enough credence has been given to any other possible system perhaps being more efficient and more adaptable. With economists and politicians spending most of their time trying to defend and match figures to suit markets in a set economic vision, the system is further complicated by individualized world-views, false beliefs, and personal interests. It has become obvious now that these biases and manipulations impair an already maladjusted system.

All systems initially unfold with much promise, and certainly when Adam Smith fathered capitalism, and when the empathetic English cleric Thomas Robert Malthus tried to match the issue of growing poverty and fertility, their intentions were honorable. Karl Marx also embarked on the noble journey of addressing inequality and unfairness. In the end all these valuable contributions, it now seems evident, were manipulated by systems to suit elect needs driven by the obsolescence of ego, insecurity and greed. And where they perhaps could all have succeeded in contributing to a more moral world, they in the end all failed, set in a society that seems to value obstinacy and conflict more than benevolence and fairness.

We should now much more enthusiastically rush to the acceptance of the plasticity of our DNA, and with life seen as an all-regarding perceptive affair in an interconnected world, we can reinforce our hope here for change. Rather than isolating ourselves and setting insensitive barriers, we can now frown upon any reductionist model with no scope to adapt to a principled and progressive society and its growing requirements.

## 4.1 Greed

We can define greed as "having or showing a selfish desire to have more of something (such as money or food)" (www.merriam-webster.com/dictionary/greed). Auspiciously for the time we live in, and to revive the much-persecuted philosophies of Karl Marx, 'under capitalism money becomes an ideal and humans become alienated in an impersonal economy that rules over human endeavor.'

Due to a much more prominent contemporary following and therefore significant influence, we can also not overlook the current Pope Francis's opinion on such matters. In his announcement of the *Evangelii Gaudium* (2013), we can clearly gather from the following extract that the Pope also has significant concerns related to our current condition and future destiny:

> In this context, some people continue to defend trickle-down theories which assume that economic growth, encouraged by a free market, will inevitably succeed in bringing about greater justice and inclusiveness in the world. This opinion, which has never been confirmed by the facts, expresses a crude and naïve trust in the goodness of those wielding economic power and in the sacralized workings of the prevailing economic system. Meanwhile, the excluded are still waiting. To sustain a lifestyle which excludes others, or to sustain enthusiasm for that selfish ideal, a globalization of indifference has developed. Almost without being aware of it, we end up being incapable of feeling compassion at the outcry of the poor, weeping for other people's pain, and feeling a need to help them, as though all this were someone else's responsibility and not our own. The culture of prosperity deadens us; we are thrilled if the

market offers us something new to purchase. In the meantime all those lives stunted for lack of opportunity seem a mere spectacle; they fail to move us.
Paragraph 54 from the *Evangelii Gaudium* (2013)

The Pope, standing on a different and more elevated pedestal for his many followers, clearly also points to a direction where wealth may fail to impress in the future, and greed more-and-more become a form of social wrongdoing—adversely affecting all of us regardless of our religious beliefs. Indications are also of a much more interconnected global society emerging, with renewed focus on the importance of morality and the environment that sustains it, rather than becoming submissive to the profits to be scavenged by human-created technology. Expressed here and with much concern, we can also perhaps, without any further discourse needed, accept that fraud, corruption, distrust and even anarchy may escalate, unless there is some concerted effort to change the current ethos of our economic structure and the healthcare it directs in the growing isolation and corruption it is getting renowned for.

With greed inevitably linked to unnecessary suffering and disparity, it is now, in its final analysis, also exposed as inarguably a despicable, injudicious and naturally unadjusted affair, deeply imbedded in our current economic model, with no positive contribution to society, environment or genome. Accusing the current fiscal system of fueling greed and not matching the needs of our moral evolution is perhaps embarrassingly direct, but undeniably getting too close to the heart of the matter to ignore any longer.

* * *

Turning then to a more thoughtful and pragmatic science for consolation, we can empirically on a primitive level accept that our assessment of need is naturalistic in origin and physiologically and genetically determined by how fast our food runs out, or by fear of us perhaps becoming a food source. Pivoting around natural events like the earth's rotation around the sun, seasons, and how long it takes for food (energy) to be assimilated before the next meal, we have, based on the cellular need for energy and an evolutionary drive to perceive and interconnect, evolved our entire primitive based fear-greed-time concept. We can then define our genetically attuned biological concept of time as the period needed before completion of certain physiological actions (in alignment with our habitat) to prevent the demise of an organism and finish a life cycle; from this also evolved our perception, including that of time, and entire epistemology.

Fear in turn can then be taken as an evolving awareness of running out of time or resources before cellular demands are met or meet their inevitable demise…in the end, time unavoidably will always take care of this much-overexpressed concern. This rather basic and superficial outlook, as explained here, is essentially driven by a need to supply nutrients and above all energy to keep the organism functioning and safe. Clearly now, we understand life is also much more than such a simple solipsistic affair to keep a single isolated cell in food and comfort or nurture its excesses while suppressing others. Subsequently, we can dissect greed down to a primeval insensitive drive to isolate resources under control of an individual or group (remember cancer cells) as non-progressive in a diminished perception and driven by an unrealistic fear (see next section). The immediate result is a manipulation of the potential flow (blood in the case of cancer cells) of resources and knowledge, and delaying or modifying knowledge to suit elect needs. Again, this reminds us of our current economic system.

We can see how such a greed/fear-driven oversimplification and neglect of an interconnected perception on all levels can hamper the natural flux of knowledge and life. It is after all a healthy body that hosts a healthy brain, and good morals that support and progress a civilized society—*not* a solitary fear-driven, greedy cell running out of time to accumulate and cling to immediate resources. In biology, such a cell will self-destruct. We can now prevent such mass destruction with a unified $\sum\infty\Delta C\{\infty\Delta a\approx\infty\Delta b\}$, open to more reliable spheres of reasoning.

It is suggested here that morality, cognition and life (to be justly defined as life) cannot be frozen in time, even for short periods, as it must constantly adapt to an interconnected progressive living network. So, greed and fear, striving to isolate resources in such a network, is not only immoral, irrefutably illogical, 'unnatural,' but also perverse in a living cosmos evolving a selfless progression of knowledge and progressive understanding.

We should note how, set in reductionist concepts of matter and objects, we can severely limit knowledge, in a Kantian sense of looking at an object (*Ding an sich*). And if it stands alone in a non-progressive Physical sphere, while not simultaneously freely evolving with the Logical and Metaphysical spheres, it is open to be corrupted by manipulation, fixations, untruths and false beliefs. Our current economic system, however unfortunate it may become for its current benefactors, and perhaps in their ignorance, again matches these criteria.

Greed, which is not only manipulative but iniquitous, runs insecurely in the Logical sphere, with significant adverse effects on our Physical sphere. In the extreme it can eradicate an entire species or sub-group and cause much suffering. It remains relentless on its insecure untimely path in creating conflict for the following reasons:

- It lures people with ever elusive promises as a false escape from its own quandary. Such imaginary comfort zones become cyclic and generate ongoing want, fear of loss, greed and power struggles. There are various levels here and this would essentially target the so-called 'comfortable middle classes' through massive marketing campaigns creating unrealistic want and consumption (the foundations of a market-driven system).
- At all levels messages are circulated in various formats, from our educational system to media, flaunting riches and certain lifestyles to be within reach of all those who perform certain actions or obtain certain objects, privileged knowledge, or make certain 'wise' investment choices; in most cases this creates

false aspirations and distractions not true to the individual, their full potential or genetic uniqueness and moral destiny.

- Consumers are hopeful and erroneously believe that this will change the current unpleasantries in their lives, but their hope then becomes obscured and mocks the essence of their *being* and creativity.
- Greed is innately linked to a fear of loss and deprivation, so it stays cyclic even in societies with excess.

Greed in its worst form is perhaps seen today in the corporate world, where mergers result in monopolies and have become a favorite 'war-game' in a business world openly associating itself with combat, with volatile markets their battlefields. Argued by some as serving the 'greater good' in having the control of 'wealth' by an elect few and that wealth will trickle down, the only truth in this statement is perhaps being mindful of its ominous aim: control. In a search for teleological justification then, people driven by such an ideology mistakenly believe, by gaining from the efforts and struggles of a 'needy' market, that they can rescue their morality and in the end perhaps act as philanthropists, setting their own version of what is 'good.' This inevitably also only happens once their own egos and needs are satisfied. Ego is hard to satisfy, and as we all know, needs differ vastly between people and can be insatiable.

This strategy is also often used as a defense by drug companies when marketing expensive drugs or supplements with overstated benefits, valiantly claiming that profits and actions are 'reinvested' in future research to benefit us all. Besides passing the tax burden back, to the now twice paying consumer, the tax benefits to corporate as a result of this are obvious. Numerous of these, at times rather feckless and overstated 'remedies,' are mass distributed. And once initial over-inflated benefits of one are revealed as insufficient, they are quickly absorbed by the next equally precarious one. The overlooked value of lifestyle improvements, healthy environments and chemical avoidance to our health are rapidly swamped by pharma and their marketing strategies and the media that promote them. The lack of wisdom and morality in all this is clear. We should also continue to question why a conglomerate with a questionable ethic, driven essentially by profits, should be able to secure its power at the cost of others where they are at their most vulnerable. In such an undemocratic manner, these corporations further gain status to then predicate our future healthcare and destiny as a species.

Perplexing in this model is the lack of openness to adequately emphasize the potential side-effects, and at times ineffectiveness, of such medicaments in some patients. In extreme the American Nutrition Association exposed such harmful effects in a publication of the *American Journal of Medicine* (July 1998). Here only one of the more familiar and commonly used groups, the nonsteroidal anti-inflammatory analgesics, or NSAIDs, were placed under the spotlight. It alarmingly revealed that NSAIDs cause over 16,000 deaths a year in the United States alone from their use in the treatment of osteoarthritis *only* (*Nutrition Digest* 38(2) [2011]). 15 million prescriptions of NSAIDs are given per year, resulting in about 30 percent of hospital admissions due to adverse drug reactions (Davis and Robson, 2016). Many doctors and researchers are urging that the drug industry should at least better prepare consumers with warnings of the potential harm as much as it promotes the potential benefits. Doctors generally try their best, but it is becoming increasingly hard in an era where drug companies openly launch powerful marketing campaigns and aggressively try to infiltrate teaching facilities, research and conferences, and on various levels exploit the intrigue and enthusiasm of young practitioners and academics. Meeting corporate standards based on economic performance has now become an added concern for the practitioner, already struggling to keep up with an increased flow of knowledge, new treatment options and patient demands. All this is further still influenced by the (in)famous 'Doctor Google' with a now much more informed public.

Furthermore, with pharmogenomics in the embryonic stages of entering the arena of drug metabolism, things will soon become even more complex here when, based on a simple genetic screen, the doctor can determine if a patient may have adverse effects to some drugs. Some genomicists are revealing marked differences in the response of patients taking drugs based on each having their own unique 'mobile' genetic make-up and individualized response, while simultaneously all these individuals can also respond differently under the influence of *diverse environments*. This will make it difficult for drug companies to justify hiding behind the term 'ethical' drugs and often inflated claims. Newly reviewed now as posing potential harm to *both* the genome of some individuals *and* the environment, the ethical implications are taken to another level. With evidence recently that this negative impact can be carried *transgenerationally* for many generations to come, the complexity and moral responsibility escalates to exceptional levels.

Michael Skinner from Washington State University in Pullman, WA, has recently shown just how devastating such epigenetic damage can be (https://doi.org/10.1038/s41598-018-23612-y [2018]). Exposing rats to traces of commonly encountered chemicals, including toxins such as the insect repellent DEET, the still-ubiquitous plastic BPA (bisphenol A), the historic insecticide DDT, and hydrocarbons such as jet fuel and oil, during a key period of fetal development, it had a genomic impact reaching as far as *ten* generations. This reflected alarmingly in the occurrence of numerous cancers, allergies and autoimmune diseases—these diseases now also proportionally more frequently encountered in general veterinary and medical practice.

The ethics of major conglomerates have been witnessed repeatedly through the last few decades, with the end result a mere swallowing up of smaller enterprises and subsequent price fixing. This ethos then affects the flow of trustworthy information and knowledge. Perhaps an extreme case was seen in recent years when the new CEO of one of these amalgamations tried to justify an increase from $13.50 to $700 on the price of *one* tablet. The morality of this is easy to question. We should query why one individual or a select sector of society should be placed on higher moral ground in deciding what is for the greater good and how it should be managed and distributed to others. It becomes even worse when many of these individuals are often not trained in the subject where they deploy their personalized views, or necessarily understand (or have a need to involve themselves with) the suffering of those affected. We simply cannot place any monopolist approach and the complex sociopolitical arena it operates in on higher moral standing in a realistic Physical sphere—and then continue to expect it, based on the genesis of what pays, to trickle down justly to those in need. Greed and attempts to control or manipulate knowledge impair our Logical sphere and make our Physical sphere obtuse. This is done, furthermore, in the light of, as already revealed, a growing immorality among those hungry for power.

Future discussions and many counter-arguments and actions may and will eventuate, but the purpose here is merely to provoke and stimulate better understanding of the urgency in addressing how our enormous potential is being hampered by unnecessary greed and an obtuse, unprincipled economic model. We shall leave this vexatious topic for now and turn to fear next—equally parlous to a sound perception.

## 4.2 Fear

*Between what is said and not meant and what is meant and not said, most of love is lost.*
Kahlil Gibran

Fear pertinaciously emanates from concern of loss. This could be fear of loss of life, family, belongings, safety, status or resources running out and so forth. Fear, besides its obvious protective value, should be acknowledged here because of its potential to also manipulate determinative exchanges between our spheres of perception. Posing a threat to sound judgment and open exchange of knowledge, we can ill afford facing complex new-era issues with a Physical sphere swayed by *unjustified* fear. Such misdirected fear is often employed in attempts to sway opinion and gain support aimed at fixing outcomes; remember the 'cookie monster.' With the prominence of topics on climate change and the impact on food and water resources and the threat of mass unemployment subject to Artificial Intelligence, we simply cannot afford to misplay fear. Think here how often fear is manipulated to serve as a marketing tool by media or propaganda by politicians.

We can also be fearful of the unfamiliar (ghosts and aliens) and unknown (unexplored territory) or novel ideas that could shake the foundations and security of our beliefs. Fear, then, is a close relative of other emotions co-evolved with perception, with the pragmatic and protective value of creating caution when faced with potential danger or shortages. Fear can, however, also easily progress into unnecessary distrust, anxiety and antagonism, with unrealistic defense and protectionist mechanisms kicking in, insidious in our society today. Such proclivity can sadly then be exploited for political reasons or by enterprises (such as pharma) to benefit from such unrealistic angst. We may relate perhaps to that 'oops, forgot to take my supplement today' anxiety. Misdirected fear is currently also leaving a significant footprint on our mental and physical health as an agent for stress, anxiety and depression; it can further boost isolation and cause insular groups to emerge. We do not have to enter tedious discourse here on this issue, but we can accept that there is an escalation in mental health problems in society today, with fear metamorphosing into anger responsible for much of this strife. Above all we should avoid a suppressed society, fearful to air opinions freely.

* * *

From a different perspective and historically, the philosopher William Champeaux (1070–1122) perhaps over-ambitiously elevated fear as the originator and driving force of all knowledge. Clearly, if an organism expresses no fear, it is more likely to come to harm and less likely to plan ahead to avoid danger. However, when this fear is not part of a sound perception, supported by trustworthy knowledge and circulating reliably between all spheres of reasoning, it is also very easily turned into an instigator of greed, anger, anxiety, false beliefs, suppression and war. Fear of loss or not having enough of something forms an unnecessary barricade against open communication, this resulting in protectionism and feud, reciprocally contracting knowledge. In our modern interconnected world with hi-tech means of communication and with all of us inevitably confronting the same delicate and urgent issues today, we can even more so now ill-afford the nugatory manipulating of information to gain support driven by unrealistic fear. Now in an era of explosive idea-making, we should avoid fear then (or knowledge) of becoming no more than a useful implement for those trying to manipulate others and influence outcomes for personal gain or profits.

In its most benign form we all know how media and pseudoscience are relentlessly trying to convince us, through clever marketing strategies, that failing to take a daily supplement (again as an example) with 'evidence-based' health benefits may make us function less well or even die prematurely. We have already seen how at its best, with the gold standard set by evidence-based research, this can be based on very shaky evidence and even dangerous when applied to individual cases. This gets even more odious when remedies such as many traditional natural remedies, with mostly no clearly proven effect on healthcare outcomes, are supported in quixotic research done by enterprises mainly directed to profit from selling these products. Mostly attempting to hide under the rickety umbrella of 'natural or holistic,' they grab at any evidence they can in support of these products to create mass sales, at times with overlooked adverse environmental impact.

Politicians furthermore can also thrive on disquietude and often obtain financial backing from large corporations to fund their campaigns, motivated by similar interests or outcomes. Gaining traction from promising resolutions to problems not clearly defined and swamped by biases, fear can be grotesquely misused here. We can see how Champeaux was perhaps correct in hinting towards fear as beneficial to drive knowledge, *only* when operating in a trustable network. Inauthentic knowledge driven by unrealistic fear, albeit still knowledge, is insulting to a panoptic perception and needs to be firmly contained in the Logical sphere or recycled to the Metaphysical. And even contained in the Logical sphere with potential value, the overall idea is still wrong when we see it as inciting unnecessary conflict based on personalized opinions and interests of a narrow group.

In attempts to segregate people, governments and political systems are at times focused on playing up potential threats, whether true or not, either to their ideologies or 'our' continued safe existence to sway outcomes. Furthermore, powerful enterprises are always hovering in the background keen to offer financial support. While controlling supply and demand of resources, be it manipulating property values, shortages, share markets or commodities, such perturbation then acts as a whip to drive a hierarchy where society and its evolving epistemology is at risk and fearful. It may even be beneficial for such enterprises to juggle employment figures and housing markets to serve their own interests. This effect has more recently been compounded, as mentioned, by 'robophobia,' now no longer belonging to the realms of science fiction, but an impending reality. With vacillating opinions, this technology is not far away from being able to replace a considerable number of our current global workforce. With the main stakeholders and owners of large corporations fighting for patent-rights on this technology and welcoming the opportunity of increasing their profits and reducing their responsibilities, many workers are in fear of losing their livelihoods. Whereas the principal stakeholders are concerned about how to remain in control of a system that has become too complex for them to understand and predict, everyone is now anxious about the outcome and clinging onto what they can. These fears clearly call for a cognition defiantly *not* to be swayed by the unrealistic phobias and anxieties of opposing groups, each contriving their own fears and biases while grabbing what they can. We have already mentioned how AI and robotics, if approached in our Physical sphere under a universal ethic, should be welcomed in the new era as a co-compatriot of our evolutionary epistemology to improve conditions for all of us.

The basic fear of not having adequate food to eat or clean water to drink can be seen as a necessary fear. The requirement for food and water as vital to our existence and health is still used to exploit the needs of the more lucrative markets in fad diets and processed foods, with a significant environmental impact. Cashing in on already over-consuming customers, there is little regard for much healthier local fresh food sources, sustainable farming techniques or the positive impact of health based on simple food choices and lifestyles. Not in full regard of the impact on the environment, it exploits rather than interconnects and subsequently forgets those truly in need of even basic nutrition, while over-servicing those already in excess. And realistically, the best diets both medicine and environmentalists now tell us, should be individualized, calorie and chemical reduced, simple and locally harvested—working closely with the environment before supplying or pleasing more exotic cravings and export markets. Export markets generally also aim at mass production to satisfy the needs of the affluent and modish; backed by lavish marketing campaigns, they focus on these more lucrative markets. The fad diet market has also become fatuous when all of science is aware that it comes down to healthier chemical-reduced vegetarian-based diets and eating less. When scantily supplemented by fish from clean oceans, and very little or no meat, and if meat is needed then humanely and sensibly farmed, this could significantly benefit both our health and moral advancement. This will inarguably also reduce much of the current health and environmental issues we face and save many sentient beings from the still needless suffering they endure. As historically with the tobacco and petrochemical industries, we can expect that logic will continue to be overshadowed here, also by what pays, with arm-wrestling between powerful figures for some years to come; change here will then soon be too late, causing significant harm and suffering to every part of this living network.

Besides such simple wisdom superficially easy to understand, chemical-ridden novelty and junk-foods, stripping of farmland, together with the harmful tobacco and alcohol markets, are still thriving while leaving significant scars on both society and its habitat. With massive amounts spent on promoting and marketing these products to the markets they furnish, and the pesticides used to sustain such mass production, ignorance and greed vastly overshadow the much-needed funding for research on sustained-farming practices, mental health, addictive behavior and environmental enrichment programs. Increasing taxes and price hikes on these products have hardly helped the already strained healthcare budgets of most governments and resulted in no more than a mere trickling back of funding for research on these major and urgent issues. This is especially concerning when compared to the massive funding available through corporations for research on novelty drugs or new technology. With numerous papers and publications on these topics and ongoing discourse, we will mention and leave these complex and important issues here for now as a future topic in spheres of perception.

* * *

With misconstrued application of resources governed by what pays rather than what is right, a growing fear of being isolated from what may become an insulated and unaffordable 'progress' is perhaps also not unrealistic. Looking at the immediate world around us (depending on who and where you are), food or land to live on and grow food is not in short supply and neither is money (governments print it) or technology. Simultaneously the world carries a ticking 160 trillion-dollar debt while continuing to distribute toxic substances and pump out pollutants in overpopulated and unhealthy cities. With large profits in interests collected, isolating themselves and secured from these growing debts and contamination, investors are constantly reminded of the 'wisdom' of their property or market investments by the main benefactors of this juggled debt with all its uncertainty.

Confronted by food waste, uninformed diet choices, excesses, starvation, poverty, while some furnish their divertissements in the same world that witnesses poverty and claims overall bankruptcy, we inevitably all face the same unpredictability, now complicated by the inescapable and real threat of environmental disasters with potential large-scale impact. How can we understand and accommodate this in any realistic and sober Physical sphere? We certainly will not make any headway in our search for morality and a dependable Physical sphere while set in such a scheme. We should ask ourselves: have fear and greed perhaps blinded us?

Affected by who we are, our education and background, beliefs, culture and exposure to this world and influenced by social media, we can formulate our own personalized views of why this is so. Our list may perhaps go something like this:

a) It is natural selection at play, selecting the fit to create genetic diversity in a rapacious society (this hopefully now sounds unsubstantial after our discussion here).
b) Greed is an inevitable sin of humankind, punishable or forgiven by an individualized religion and its deity as set; based on culture and heritage, we should also help the needy and pray for them while defending this set belief. Perhaps we can more objectively claim (without being blasphemous) that this group is over-endowed with monoamine transporter 2 (VMAT2) that predisposes humans toward spiritual or mystic experiences, the so-called 'God gene' as suggested by some scientists.
c) The rogue approach: the poor or underprivileged are an ignorant lot, incompetent and slow to adapt to knowledge and it is generally their own doing. The impact of drug addiction, poor diet, lack of education or alcohol abuse is also their weakness or bad luck in a tough old world. Still pretty smart, though, if you can scavenge profits off such an addictive and disconcerting bunch; it may even help in their early exclusion in an overpopulated world. The sociopath gene is evident here.
d) The vacuous 'it is just the way things are and always will be with genetic diversity and natural selection' (a sadly popular view among some 'educated' comfortable classes).
e) Life is unfair…a favorite among those vaguely familiar with the utilitarian value of the metaphysical but not inclined to religiousness.

f) The system has failed—blame the government, a favorite among despondent workers in developed economies.
g) Conspiracy theories. Those theories that the 'conspiracists' (whoever they may be) are constantly trying to suppress.
h) *Or* perhaps more sensibly after reading this book, some may now see all the above as not reliably interconnecting between all spheres of reasoning, and as a manipulation of the Logical sphere (full of biases, greed and fear) afflicting the soundness of the Physical sphere.

In view of my argument so far and keeping in mind the respect for an interconnected network of life, I will not attempt to defend or discuss any further the nebulous points raised under (a) to (g). Our focus is now on (h) and the compelling question is raised: why are our Logical spheres so easily manipulated by false beliefs and illogical deductions and lured and swayed by profits?

In the famed awareness-test run by the psychologists Daniel Simons and Christopher Chabris at Harvard University in 1997, students dressed in black and others in white passed basketballs between the participants. When participants were asked to count how many times the basketballs were circulated between the participants dressed in white, only 30 percent of the participants in conclusion of this little experiment saw an individual dressed in a gorilla-suit cross the screen. What this experiment proved is what is referred to by psychologists as *selection attention deficit.*

This experiment can be viewed at https://www.youtube.com/watch?v=vJG698U2Mvo.

In brief, what it accounts for is that we are prone to the curse of seeing what we are conditioned to look for, and we may often miss other salient facts based on our conditioning. If I (as an example) was trained to spot gorilla-suit clad individuals, I would most likely instantly spot the gorilla. With our biases in turn affected, as discussed in this text (by many individually determined factors depending on who, what and where), we can see how easy it is to create world-views based on personal biases and be subject to dominating input and reigning belief.

We all know that a few people may profit significantly as promoted by the main benefactors and supporters of the shares-market concept. Large sums of money are daily circulated, based on abstract values and 'calculated predictions' around the availability of commodities. This can be significantly influenced by supply and demand and the biases of those with vested interests in these outcomes and enough power to sway, control or influence political outcomes that may affect these markets. From international currencies, the price of gold, oil, and various food and meat commodities, many may gain or lose in these conjectures. Driven by the lure of investment value, such uncertainty is forced to promise security to the hopeful investor; we were all once wisely told that gambling is a bad habit. Many other similar commodities 'float' on shares markets with questionable benefits and exposed to many potential biases. Most of these large sums of money are hoarded by larger conglomerates at the expense or loss of smaller investors or the producers. The producers are often farmers struggling while supplying our vital needs. Such vacuous figures, constantly shifting between accounts, are essentially abstract and, as mentioned, can be strongly influenced by global politics, elect biases and personal worldviews, all trying to stage confidence in much precariousness.

Adding to this vexation of insecure markets that our economic structure is structured around, according to some research, the executive with sociopathic traits is also common in the corporate world today, although most such individuals will not typically wind up in prison (seen as stealing from the people). They have also been referred to as "socialized psychopaths or sociopaths" by some researchers (Kent, K.A., Buckholtz, J.W. 'Inside the mind of a psychopath.' *Scientific American Mind and Brain*, pp22–6 [Oct/Sept 2010]). This article raises concern that, "In fact, many are promoted explicitly due to callousness and ruthlessness they demonstrate and wind up in the cushioned leather chairs of the executive office in a society that blindly rewards profit rather than moral conduct." Aiming for a moral society relying on individuals with opposing traits of ethical behavior, we can see the possibility of power being linked to a growing immorality where those with disregard for others often tend to get better rewarded. This genetic drift of an over-presentation, torn between a select God-gene on the one side and a sociopath-reward system on the other, may then become a significant threat to a civilized society—a large-scale selective attention deficit or *social attention deficit*, with a disastrous outcome for us all.

* * *

The question remains: Why do we allow such easy manipulation of our Logical sphere to influence outcomes in our Physical sphere based on unproven ideas, uncertainty and seemingly unsettled social hierarchies and genetic traits? A simple answer may be hidden in the cyclic activity of such a fear/greed-driven affair, creating few options for us but to create new ideas based on the very nature of what a fear/greed interaction proselytizes. Subsequently, all is drowned in this social attention deficit created by domineering opinion, whether such opinion is true or false (more on this under the section on mentalizing).

We can see how the subjectivity of both greed and fear can be used to manipulate the progression of emerging knowledge and affect all spheres of our perception. A society (or individual) failing to eradicate at all costs the corruption and social attention deficits afflicting its evolving understanding and knowledge, cannot operate in a healthy interchange between all three spheres of perception, and will not be able to realistically confront more urgent future concerns while calling itself civilized. Such reserve and attention deficit, fueling fear and greed, whether done in ignorance, tradition, or through a growing over-presentation of some genetic trait, while causing manipulation of supply and demand of resources or knowledge to suit an elect group, can now be seen as self-limiting, dangerous and inarguably unethical. We have already postulated how trying to justify this under a traditional Darwinism and reductionist natural selection concept will simply no longer suffice. Clearly then, and now with better understanding of the evolutionary process, we certainly are wiser and better off searching for reasons to advance benevolence, fairness and altruism in a progressive perceptive network, rather than animus in egocentricism and isolation.

## 4.3 Beauty

*Beauty is the promise of happiness.*
Stendhal

When we think of describing beauty, we may be inclined to use a conjecture based on lines, shape, form, and color, in recollection perhaps of an object, a person, or a place. But surely there must be more to explain that feeling of euphoria and veneration when we experience unblemished beauty? We should also not neglect beauty while focused on the adversity of fear and greed and overlook this as perhaps the most righteous achievement of a perceptive evolution.

This is another area where we may show *severe* attention deficit in today's world. So easily distracted by phobias and biases in a society inclined to play up fear and superficial appearances, we often fail to see the beauty of our world and in others while swamped by materialism and objectivity.

* * *

The historically argued proximity of goodness (morality), beauty, and truth goes back as far as Pythagorean times and has been critically argued by most philosophers ever since. In our new understanding, we should perhaps again pay tribute to this Greek philosopher and his insight from twenty-four centuries ago. Assisting our evolving epistemology and our search for understanding beauty with his contributions to mathematics, architecture, and science, he also caused ongoing controversy by claiming a certain consonance exists between the 'human souls' in accord with the universe and melodious musical notes, architectural shapes, forms, and mathematical equations. He believed such harmony brought us closer to a fusion of these three elements of beauty, goodness, and truth.

Few instances come to mind where moral action generally is not backed by goodness or where goodness does not relate to moral action and how this can open us to the beauty of life. This Pythagorean view, however, reaches much deeper and becomes more intelligible when we tie it together with current understanding of our evolutionary origins. We now understand how we cannot separate our morality from our perception. From this unison also emerges the only means available to us for sensing and experiencing beauty.

With such harmonious interconnections now evident in a better understood and interconnected biological world, operating in synchrony with its environment, perhaps Pythagoras's concept was not as far-fetched or strange as it may appear at first. Typifying the quest of the philosopher and in constant search to better relate to the physical world, he was striving to connect the metaphysical and mathematical (material) worlds. Inarguably, some unconventional thinking and profound insight were required to sense beyond reducing the world to mere set objects and numbers (this inadequacy now also realized in modern quantum physics). Now aware of a principled interactive molecular evolution, and after being enlightened by Darwin, the DNA helix, and other recent discoveries in the sciences, it is much easier for us to comprehend the accord needed to sprout the interconnected complexity and consonance we witness in our changing world and universe. Such continuity explains the evolution of things such as galaxies, flowers, and a human brain with its ability to bring joy through the arts, music, and life. First reducing this organic perceptive organ down to molecules, we are again now amazed by this molecular arrangement and its ability to give us this wondrous experience that so liberated our perception.

The rather superficial argument often used against such a Pythagorean world, so typical of the reductionist era, runs along the lines of: "If I listen to euphonious music with rhythms attuned with the motions of the universe, I do not as a result necessarily become a good person and can then lay claim to sensing the truth." This can comfortably be seen as misconstrued and not delving deep enough into what he may have reached for.

Important to us here, however, is how Pythagoras already sensed consonance instead of disparity between the structure of the human 'soul,' our ability to create, and the structure of the cosmos. He was perhaps prodigiously aware of how a certain cognate morality (intertwined normative) was needed for such objective interactions in an interconnected universe to make it function the way science is now deciphering it for us today, and we can employ it in our three spheres of perception. And when notes harmonize to form melodious music, or when beautiful art or prose is produced, and disparity is reduced in a strife-torn society, it can only be proximations of perception *and* morality. This result inevitably must be termed *beauty*. We can also revive this Pythagorean view with our proposed formula here, where we see a certain mellifluence in the ethos needed for such harmonious interaction in a highly principled perceptive evolution as:

$$Ev(mo)=\sum\infty\Delta C\{\infty\Delta a(\text{Metaphysical}\rightleftharpoons LSR\rightleftharpoons PSR) \approx \infty\Delta b(\text{Metaphysical}\rightleftharpoons LSR\rightleftharpoons PSR)\}$$

This certainly again does not imply that if I listen to Mozart's *Requiem* or grasp what the above formula implies, I will become a good person or suddenly see the truth, or necessarily sense the beauty in this composition. It does, however, suggest that Mozart managed to align these elements in the observer and the observed. Subsequently, such a composition may liberate our perception and bring us *closer* to those experiences where truth, goodness (morality) and beauty meet and result in moments of joy. But only when all three spheres of perception interact, interconnect and are harmonized in knowledge and understanding reflecting workable truths, be it harmonious music, mathematics, a painting or interactions between molecules, will *a* harmonize with *b* and result *in ultimate beauty*.

Even the threat of a supernova explosion, seen in the context of this understanding where universal forces follow certain physical laws based on harmonious interchanges, now becomes a natural event with an element of harmony and beauty. All metamorphosizes in a universal perception aiming for *principled* interaction and consonance in persistent change between all spheres of perception. Without understanding and realization of such an interactive world and universe following interdependent principled laws while harmonizing with change, we may fail to gain understanding of the morality it is part of and see it as a threat and therefore fail while operating in isolation. The better harmonized and the more reliably our three spheres of perception are interacting with change, the closer we can also get to this morality and its infinite beauty, and the less fearful the world and universe become.

* * *

Other examples may be taken here, such as the superficial beauty in a piece of art or a face in a crowd, where the beauty *may* analytically conform to certain lines and proportions of the painting or the face. But it is only when we *see* and *understand* it as part of an interconnected whole that has the potential to connect us with purity, goodness and truth that it can become less alienated. Again, this certainly does not suggest that the beautiful face in the crowd belongs to an honest and good person. The beauty of the face, eyes or conformity of an artwork merely represents a small part in hinting to these elements of morality we are so critically trying to harmonize in ourselves and the world around us. Elements that, only once all three are combined in both *a* and *b*, will result in more than an incomplete 'beauty.'

In the infrequency of this alignment also lies the ephemerality of beauty.

An essential element of beauty then is to not only acknowledge it as part of an interconnected whole but as constantly changing and ephemeral. Seen in isolation and not transitory, it cannot unfold its full potential but drowns in its own ego. We can subsequently also never experience beauty by claiming possession. This is because of the egocentrism that transgresses possession, where we immediately destroy morality and goodness, the vital ingredients of the elements we attempt to interconnect and harmonize. And clearly there is no goodness or morality in isolating beauty from others, *or* stagnating change. The latter especially applies to alluring memories we so desperately attempt to replicate. That search for the same exaltation experienced years back while visiting the Colosseum, with the sun setting in the background, may not be the same on a repeat visit to Rome when confronted by a queue of sweating tourists in the sweltering midday sun.

Once the beauty of an object or person *b* is realized and synchronized with these elements in the observer *a*, beauty is now in both the eye of the beholder and the observed. The synchrony of these elements between observer *a* and the observed *b* is perhaps what Pythagoras also aimed at in making these connections. With deeper digging and better understanding of the constant change and interconnection between the observer and the observed, it can be seen how we can merely appreciate, respect, and enjoy beauty (or be joyful about the prospect of encountering it again) but never constrain or possess it. We can merely strive to harmonize ourselves with beauty, truth, and goodness, and for a few happy moments, through constantly improving ourselves by aligning these elements in ourselves, *more* frequently experience it. From this we can gather that beauty is not only simultaneously elusive and changing all the time but also more compelling and ubiquitous (and part of us) than we tend to think. When reliably interconnected in all three spheres of perception, we may also now experience more of these entrancing moments without attachment and filled with joy.

The potential of beauty then harbors itself in this search for such harmony within ourselves for these three evanescent elements. It is in *understanding and developing this longing for goodness and morality in our actions, free of possession or obsession,* that we can also increase the frequency of these exposures to beauty and realize we can never entirely dominate it as we *are* part of it. This is also where we may find our own morality. This disparate attempt at bringing us closer to beauty through possession, linked to fear and greed, is exactly what then also estranges us from it.

We may sense a similarity here between the ephemerality of beauty and that of obtaining ultimate knowledge. The more freely and perspicaciously we move between objects and experiences and share our knowledge without fixations but aim at improving our understanding to interconnect these elements, the more we can evolve reliable and interconnected spheres of perception. Subsequently, functioning in such a progressive perception operating both intrinsically and extrinsically, the more beautiful and truthful our knowledge and morality become, and the more *significant* and less fearful our minuscule place in a vast and wondrous universe suddenly appears. Our perception now appears boundless.

The latter is important to many of us today where we seem to be anxiously driven to accumulate more than we need or can consume. In a consumerist society, driven by markets constantly promoting possession as a key to happiness and beauty, simultaneously in preparation of shortages or doomsday, we remain needy and fearful, even in excess. We can see how, in an attempt to obtain happiness through status or isolating wealth, we may perhaps overlook beauty as it unfolds around us, while attached to one fixed element. Interconnecting and operating in all three spheres of perception in turn relieves us of some of the environmental and social burdens that come with excesses in today's world.

We now see the beauty of ageing and realize the ugliness of our actions in isolating ourselves in egocentricity while overlooking the neglect of the environment and those in need. Sadly, many of the attempts by more pragmatic futurists and scientists today to create awareness of such superfluity are also quickly swamped by media, self-interest, sales, and profits, rather than pursued and seen as a united attempt to harmonize truth and goodness in a search of beauty and a better world we can all share. Such nescience also robs us of our cognitive uniqueness (what makes us us) and causes us to over-admire or attach ourselves to objects or certain ideas about objects, or expect epiphany in one specific event. If we continue with such a myopic view, that event could soon become the extinction of both our species and the beauty we all strive for.

* * *

It is apt to end this section and reflect on the experiences of beautiful minds, even in the face of death. We can perhaps take a 'light-hearted' musical break and, if possible, listen to one, if not all, of the following near-death compositions of some famous composers. This is from a quick Google search, and everyone will perhaps favor their own here. Mozart's Requiem, Bela Bartok's Piano Concerto No. 3, Schubert's *Winterreise*, and Pergolesi's *Stabat Mater* were all written near the composers' deathbeds and evoke a sense of the beauty of life and its continuity—even in death. It is perhaps Gustav Mahler's *Das Lied von der Erde*, where the composer's selfless search for consonance and beauty continued right to the end and just before he died, that best reflects the unison of life and death. The harmony and continuity of the world, life, and its infinite beauty is expressed in harmonious notes and the repetitive words at the end: *Ewig, ewig, ewig, ewig*—forever, forever, forever, forever...

We can now also see the endless beauty, truth, change, interconnection, and the transgenerational continuity carried in our 'perceptive' genes, rather than finality in even death; all we need is to combine our spheres of perception within a universal morality.

# 5

# The Impact of False Belief Systems, Mentalizing, and Culture on an Evolutionary Morality

*Sccience without religion is lame; religion without science is blind.*
Albert Einstein

*Just as a candle cannot burn without fire, so human beings cannot live without a spiritual life.*
Buddha

Culture, belief systems and religion are considered 'extra-somatic' processes by most cognitive scientists. Still ineluctably products of an organic evolution, they have the potential to heavily impact on our concepts of the world and how we share it. We suggested a progressive Physical sphere of reasoning where workable ideas are interactively driven to continuously advance a progressive perceptive evolution, formulating ideas in search of a better more functional world. We further argued the mandatory need for such a progressive perception to evolve in unison with a Logical sphere of reasoning while confronting an enigmatic Metaphysical. This in turn necessitates the need for an inimitable ethic to have truthful knowledge continuously evolve in a reliable and yet pliable Physical sphere.

Such progression, then, is constantly perceptive and responsive to changes as part of an expansive network. And on this basis, we evolve our understanding, survive, and improve the productivity and quality of our lives in a concordant cognition, constantly changing ideas about ideas. We have clearly argued why a reductionist evolutionary theory based on genetic change in *a*, occurring due to the selective suitability of genetic recruits that *happen* to fit requirements of the environment *b*, stands alone when stated as, $Ev=\Delta a(C)\rightleftharpoons\Delta b$, and will no longer suffice. All along we were hindered by such an oversimplified disjointed idea of an evolution mechanically driven to reproduce and survive, seemingly with no purpose. In this disjuncture of our own ability to be perceptive, we have failed to see cognition as only able to function when part of a principled perceptive network. Such an unavailing evolution made the alternatives of hiding behind nebulous false belief systems set in the Metaphysical equally attractive. This isolation also did not comfortably explain the complexity and diversity seen around us appearing, in geological terms, very rapidly. Now understanding evolution as pliable, progressive, interactive and perceptive, with an unobtainable end, we have placed reproduction and survival where they rightfully belong, as mere implements in advancing a progressive understanding and sublime perception. This process perpetually evolves the Metaphysical into a progressive Physical sphere and ongoing falsification in an evolving cognition. The laws of our physics and mathematics may serve to explain much of how we currently perceive objects and our relationship to them in our Physical sphere, but as our network expands and grows in its own complexity and understanding, they will very likely not remain unfalsifiable, or apply everywhere in the universe, and neither would our perception of how things interconnect not dramatically change in time.

We now propose a more functional and open formula in evolutionary biology. Facing a promising future and simultaneously carrying greater responsibility in evolving cognitive life as an *interconnected moral concern*, with full regard for all its components, we grow in complexity and perception:

$$Ev(mo)=\sum\infty\Delta C\{\infty\Delta a(\text{Metaphysical}\rightleftharpoons \text{LSR}\rightleftharpoons \text{PSR}) \approx \infty\Delta b(\text{Metaphysical}\rightleftharpoons \text{LSR}\rightleftharpoons \text{PSR})\}$$

This formula, acknowledging the importance of a sapient perception in concordance with progressive understanding and morality, can also have its sagacity afflicted by numerous *disruptions*—disease, inadequate diet, pharmaceuticals, drugs, toxins, hallucinogenic plants on an individual or collective level. And, on a larger scale, it can be affected by belief systems, culture, religion, and toxic environments. All have the potential to influence our Logical sphere of reasoning either directly or indirectly. The effect of this on any level of an integrated perception can be very significant.

Healthy relationship and interactions between all spheres of perception is conducive to best outcomes for a judicious evolution. With our morality and spiritual knowledge seated in the Metaphysical, this process is circular, as it were, and reciprocal. Moral (truthful) action and prudent exchanges between our Physical and Logical spheres make possible a free accession of pure knowledge and allow new ideas to emanate from the Metaphysical. And the accession of truthful knowledge circulating between concordant spheres makes possible the performance of altruism, which in turn enhances our capacity to share knowledge and improve our morality, and so advance as a principled progressive network. Our progressive perception, presented as a $\sum\infty\Delta C$ (unmeasurable wisdom), advances higher the higher the morality, and the more reliable and healthier these interactions, the higher both cognition and morality, $Ev(mo)=\sum\infty\Delta C$.

## 5.1 Mentalizing, Coherentists, and the Curse of Knowledge

> *The only people for me are the mad ones, the ones who are mad to live, mad to think, mad to talk, mad to be saved, desirous of everything at the same time, the ones who never yawn or say a commonplace thing, but burn, burn, burn like fabulous yellow roman candles exploding like spiders across the stars.*
> Jack Kerouac

These tantalizing words by Jack Kerouac are apt to introduce the term *mentalizing* that has recently become important in neuroscience and psychology. This involves the study of *how we form beliefs and refers to the ability to detect (assume) others' mental states or beliefs.* Mentalizing ('theory of mind') is often broken down into a set of sub-abilities, such as the more objective processing of belief-contents or decoding mental states versus more subjective reasoning about goals, ambitions and actions. Some psychologists, such as Alan Leslie et al. (2004), have suggested a 'true-belief default' where usually we guess somebody else's belief to be the same or at least similar to our own. Some researchers have referred to this cognitive bias as the 'curse of knowledge' where we simply assume things because of reigning beliefs. With both **selective attention deficit** and **true-belief default**, bias is revealed as inclined to be focused on salient facts and can clearly cause error in judgment (disjunction caused by generalization). Recent data backed by advances in neuroimaging suggest that improvements in inhibitory control create the ability, in later development, to impede the tendency to focus on salience only. This potential opens new doors to perception and the ability to be more creative and expand on more reliable ideas as we age. This potential, however, also depends on cognitive health and recognition and receptiveness to a progression of new input.

Assuming something to be true based on other beliefs or arriving at estimates based on another's belief may have some evolutionary value where quick assessments must be made in social groups, as for example in 'fight or flight' situations. Think of a herd of deer when one in the group senses the smell of a predator. Perhaps this is why inhibitory control is not developed in the young until their world is more experienced (around the age of four years in children). But such assumptions can also make our cognition arrive at lasting illogical conclusions. There is continuing evidence that neuroscience, especially backed by constantly improving neuroimaging techniques and technology, can provide us with more answers in this emerging field. Now with commercial interest in robotics as a stimulant, as mentioned in previous sections, it has also taken a new turn.

Backed by such objective data we can now confidently say that various aspects of social information are seated anatomically in diverse parts of the brain to function in a social network. Not delving into anatomical complexity more than of value to us here, the critical fact remains that healthy *interconnection*s are primary in maintaining an adjustable complex network. It is through this intercommunication between anatomically diverse brain parts that we gain the ability to override a dominant response or ignore irrelevant information not critical to our everyday life and make informed choices. Important yet again is the integration of this system with other parts and the environment, and the 'senselessness' of operating in isolation. The inability for any part to act in isolation and the potential effect of **extra-somatic manipulation** of inhibitory control is important here. If any part is non-perceptive, non-pliable and *not* part of an interconnected *reliable* network, operating in such desolation it will lose its appeal and value for confronting progressive environments and continuous change. This effect then also creates scope to dominate and delude both the Logical and Physical spheres of reasoning and implant a false belief or concept in an insulated sphere of our perception. In contrast, the more all-regarding and principled this intercommunication, the more expansive and insightful the perception.

It is also in these intricate anatomical parts where our idea-making takes place and where medically today we understand how miscommunication between these elect neuron groups can cause an array of mental diseases. It is also in these delicate interchanges where belief systems, culture or numerous drugs can affect inhibitory control, creating delusional conclusions.

Today numerous drugs have been formulated with the ability to alter these interconnections on the transmitter level, and more and more drugs and methods are likely to emerge with our better understanding of such physiological and biochemical interconnections in the brain.

It is also here where reductionism is hard at play to create robots with human traits, inevitably aimed at impeding any evolutionary progress in overcoming this inhibitory trait to focus on salient fact, to avoid the complexity of such an enormous task. We can perhaps begin to see how a religion, a belief system, drug, toxin, or culture (and soon robots) can affect an interconnection or exploit an inhibitory response of the brain and even explain, in extreme cases, the large-scale killings or mass suicides by followers of a belief system. Soberly now in recent times, we should add the growing fear of a domineering Artificial Intelligence and an emerging robophobia. In our creativity, the very thing that makes us human and gives us the ability to escape the danger of reductionism, it now appears may also be seated a *weakness* that can be exploited.

At this point in our understanding it is hard to *not* accept the organic origins of our brain, and the transgenerational perceptive evolution with all its interconnections, as giving us the ability to interact with a changing and challenging external world. The source of all our thoughts and ideas that makes us what we are, this perceptive evolution constantly drives the progressive 'idea-making' process, interconnected to past, present, and future (time). With such knowledge we now also realize that we can improve both our physical and mental health by wisely interconnecting to our environment and its contents. Should we consider the more unrealistic metaphysical possibility of an external force conducting our thoughts and actions, commanding the somatic brain, both the potential and internalism of this perceptive drive and the morality of its interconnections are lost in a reductionist design that would not have resulted in life, or continue to support it. This external force can also with relative ease, then, soon be replaced by an Artificial Intelligence. We can sense the absolute importance of *morality* to enable any system to reliably interact in all three spheres of perception for successful outcomes.

If we turn to logic just for a moment here with mentalizing now in mind, classical knowledge of propositions state: *S* (or as discussed in this text *a*) knows that *p* (or *b*), where *S* stands for the subject who has knowledge and *p* for the proposition that is known. The question then results: What are the necessary and sufficient conditions for *S* to know (mentalize) *p*? We have already proposed that this negation takes place in the Logical sphere of reasoning (anatomical regions where this is appraised, aside here) and is variable depending on what, who, where, when, and how you are, and can now add what *means* are available for assessment when *S* is trying to know *p*.

Our classification as proposed here of progressing knowledge and idea-forming, besides addressing the issue of conditions for *S* to know that *p*, also further incites the criticism to which both foundationalists (claiming that beliefs are acceptable if they cohere with fundamental beliefs) and coherentists (claiming that for beliefs to be coherent, they have to cohere with other beliefs) are exposed.

Colin McGinn, a professor at Harvard University studying behavior aspects related to business and law, expressed the weakness of coherentists as follows:

> Coherentists are committed to the view that, for example: 'Snow falls from the sky' is true if the belief that snow falls from the sky coheres with other beliefs. It follows from this and the redundancy biconditional (*p* is true if *p*) that snow falls from the sky if the belief that snow falls from the sky coheres with other beliefs. It appears then that the coherence theorist is committed to the view that snow could not fall from the sky unless the belief that snow falls from the sky coheres with other beliefs. From this it follows that things depend on what is believed about them.
> Colin McGinn, 'The Truth about Truth,' in *What Is Truth?* (ed. R. Schantz, 2002, pp194–204)

This seemed strange to McGinn and also challenging (if not derisory) to us in our attempt to present pragmatic ideas to a prodigious Physical sphere of reasoning, since snow could fall from the sky even if there were no beliefs or disbeliefs about snow, or even means to perceive such an event.

Awareness of falling snow can, however, only be perceived and interpreted with perceptive methods having evolved to do so. And this then also is sculptured around knowledge exposure and an evolutionary epistemology. The circular interchanges between our Physical, Logical, and Metaphysical spheres in turn are furthermore variable and valuated depending on who, what, and where you are. A simple belief system without evidence cannot harbor in the Physical sphere more than perhaps momentarily before its rejection. We can perhaps sense some social value in our ability to mentalize, but much more clearly can we now sense the danger if this is set on a deluded and manipulated sphere of reasoning. Originating without sufficient evidence and subsequently forced or manipulated as a *set* (false) belief, any perception is poorly equipped to confront the challenges of a new era. In an era already deeply imbedded in an advancing technology, we also urgently need improved understanding about ourselves and the changing concept of our place in an evolving universe.

A tree, a blind man in a house, or a cat in the tree will all create different versions of the same fact in their Physical sphere. The blind man inside the house may perhaps conclude that snow is falling from the sky because a good friend told him so, or if living in isolation when rubbing his cat returning from outside feeling the familiarity of snow on its fur, but this is still open to some error—the friend suffering from hallucinations or being blind-drunk as an example. The tree in turn has different perceptive means but can still be 'aware' of snow falling from the sky on a physiological level. The latter idea is not as crazy as it seems anymore, with botanists recently recording various physiological changes and communication methods in plants in response to changes in their surroundings.

We can now perhaps see how the basic justification based on experience in foundationalism can be avoided. As an example:

With the proposal here based on simple logic accommodating changeability, falsification, plasticity, and conductive inference, it would be feasible, if the weather report testifies it to be snowing on the ski-fields in Queenstown, New Zealand in July (winter there), to place this as a strong idea (belief) close to the core of our Logical sphere. We can, however, only place it in our Physical sphere if we could simultaneously transfer ourselves to the scene and experience the snow falling from the sky with our own unafflicted senses and perception. Witnessing it on a reliable electronic device such as a smartphone while talking to a friend at the ski resort, it may be strongly reliable and now enter our Physical sphere, but remain subject to time, error, or even the friend perhaps playing a prank on us—with our belief still set on this device. If my smartphone (S), however, informed me it was snowing in Fiji, I would rationally question either the source of my information, the device itself, or consider the chance of freak weather patterns very critically in my Logical sphere—well removed from the core of my Logical sphere. See Figure 2 below.

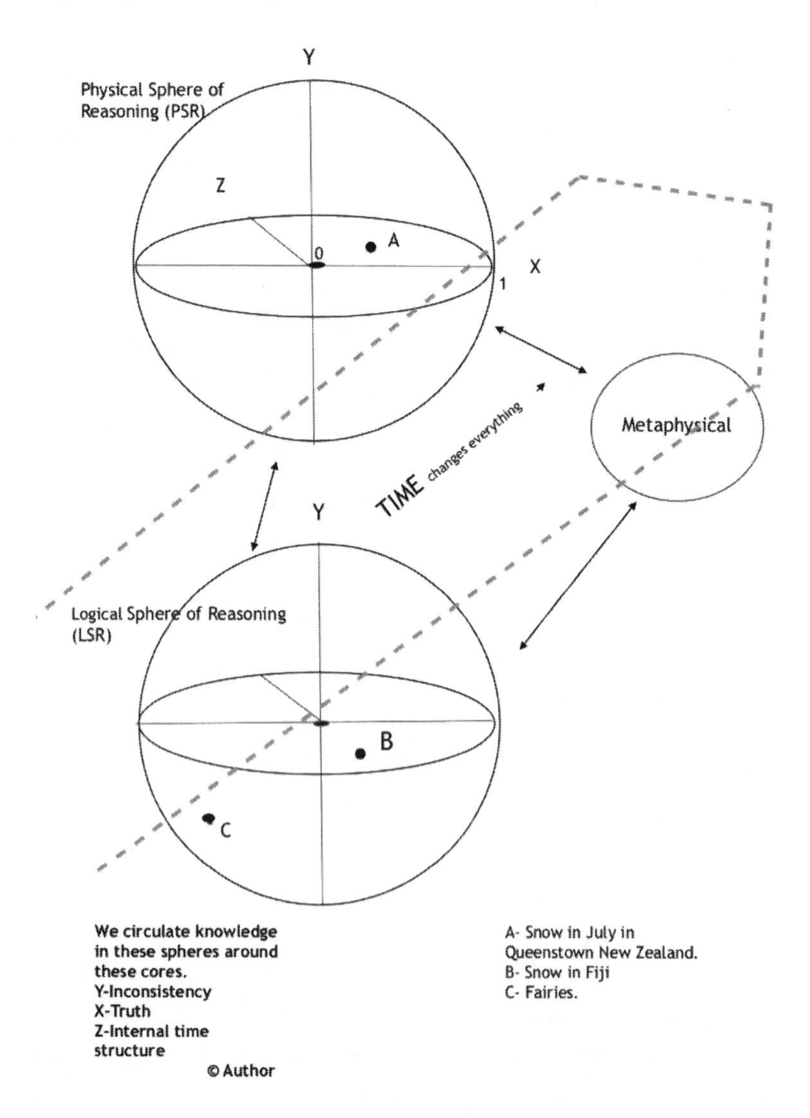

The fact remains that our evolutionarily acquired perceptive abilities gather information about our environment and, based on complex physiological interplay between memory, reason and *trust* while evolving knowledge, we can make an introspected decision to gather our skis and where to go. Other beliefs and opinions about snow in Queenstown can now circulate on social media, or whatever other means, with some predominate pattern. We will still only know if it snows when we get there when we get there, and then it could change (stop snowing) as we get there. Blind acceptance based on belief only, in ignorance of time and change, and without reliably interconnecting dependable knowledge as part of a complex network, perhaps may have some social value to maintain group order or in the instinctive 'fight or flight' reaction. However, when it comes to sober decision-making, expecting pragmatic outcomes to experience snow with *absolute* certainty, none whatsoever.

* * *

Ideas, even when tested against testimony, cannot satisfy a perspicuous Physical sphere of reasoning and will remain to create rivalry and confusion in our Logical sphere of reasoning and be open to manipulation and dispute. Here false belief systems, misinformation, drugs and mental health issues can hijack pragmatic conclusions in the Logical sphere and create the Physical sphere to turn metaphysical beliefs into ongoing conflict and needless suffering.

With belief systems such as a belief in a God, fairies, or snow in Fiji existing in our Logical sphere, with no objective evidence as support and also no means of disproving it but with enormous social impact, we simply cannot place such beliefs in our Physical sphere. The difference is that we can fly to Fiji and cross-check, while in the case of fairies we cannot. Such beliefs therefore cannot be objectively entertained in our Physical sphere but also cannot be rejected by the Logical and Metaphysical spheres. External moralism is also not an option to justify such false beliefs as we have discussed—a universal progressive moral particularism, however, has much more appeal.

We cannot help but think here of all the religious wars and aimless deaths caused by and driven by belief systems with no evidence other than believing *S* to have knowledge of *p*. The next obvious question then is *why*, if there is a patterned need to enhance cognition, and increase quality of life and survival, do such belief systems remain in place for such prolonged periods in our evolutionary cognition? We have discussed mentalizing and the curse of knowledge above and also mentioned the social value in acting to maintain group order.

Returning now to *S* knows that *p*, where *S* stands for the subject who has knowledge, and *p* for the proposition that is known, we can propose some arguments around this and why such belief systems remain in place (the reader may add to this or create his or her own list).

1. False beliefs supported by mentalizing as an immediate block to infinite possibilities of *S1*, *S2…ad infinitum*. We believe in *p* because *S* said so, to reduce confusing and conflicting ideas. This easy way out can be related to the proposed juvenile relation to the 'candy in the cupboard' scenario used by psychologists. Here the lack of cognitive inhibitory ability, as seen in children under four, prevents focus on where the knowledge originated from as perhaps contributing to where the candy can be. In turn this changes (or not) with age, as inhibitory control is improved to *not* only focus on salient facts or set beliefs but to consider other options and origins of beliefs.
2. A *utilitarian value* belief. A value belief is a belief expressing itself as of utilitarian value to the group or the individual. As an example, going to church as an atheist with the subconscious motive of supporting social or business networks.
3. A joint belief (whether true or false) can act as a social glue and it creates opportunity for moral rules and a normative to be set for such a group and divagated to a higher control. It redirects human dogma to a metaphysical dogma. It is still, however, open to manipulation (perhaps more easily) and forceful

application in a social hierarchy.

4. Such belief systems are kept in place because of an innate evolutionary need to believe in something higher and benevolent (the so-called 'God gene' as mentioned and recently described by some genetic scientists). Besides acting as a social adhesive under a paternity, it simultaneously creates hope and reason to endure the unfairness and harsh realities of our uncertain objective world and delivers some security.
5. Morality and ethical conduct are essential for societies to evolve, remain interconnected, and progress; belief systems are a means of creating a higher authority than human beings as an aid to apply such rules.

Inarguably, now we can see the innate need to trial different methods to enhance interconnection, with tolerance and morals as vital requirements in creating a veritable Physical sphere. If anything, this agnostic approach reemphasizes the need for morality. To illustrate with an everyday example, a person drinks water to keep physiological actions taking place in order to remain alive. We believe that if we are served water in a restaurant, it is safe to drink. We know and have strongly argued that cognition cannot progress without moral interconnection. It is also difficult to refute that a society without morality will not make much headway in remaining functionally interconnected and evolving intellectually—similar to a living cell in isolation or the long-term prospect of cancer cells in disregard of its host. Religion and belief systems can be respected here as ingrained attempts at creating social order and serving perhaps as an oracle to guide moral values. But inevitably operating in secular groups and 'set' in the metaphysical under dogma, they cause unnecessary segregation and conflict. We can perhaps see this as a utilitarian approach to a certain belief system, where if a religion serves as a social glue it may enforce moral values. We cannot, however, soberly do so because there is no evidence or sound justification of the deity that the belief is centered around. This becomes even more concerning if such a belief is clashing with many others and their personal beliefs, with both sides manipulating ideas to suit their set concepts. With such beliefs set on equivocal foundations in the metaphysical, we can see how we will struggle to accommodate such beliefs, even in the outer realms of the Logical sphere.

We may, however, consider such a belief in our Logical sphere as long as we realize that it is still disjointed with the Physical sphere. And how can we base an ethic and social order, reliant on evolving a dependable epistemology, based on questionable beliefs manipulated by conflicting authorities? Furthermore, once the ambivalence of such a false belief is exposed or suspected, its supporters are more likely to turn to corruption or violence and distrust morality overall, or even attack morality. This can also be seen as a worrying trend in society today where many find escape in technology and easily sidestep any topic on morality.

However, when beliefs and morals evolve and circulate between principled spheres of perception, we can respect such a morality not as dogma or metaphysical, but as a benevolent and pragmatic need to help drive a civilized society striving for proximation with an expansive cognition. This trust then goes beyond authoritarianism hiding behind a metaphysical belief searching for divine actions, but is based on ethical and moral conduct as an essential evolutionary need to progress in a trustworthy epistemology facing a united destiny. The essential relationship between morality and interaction between reliable spheres of perception and spiritual knowledge becomes circular, inevitable, and reciprocal. Moral interaction then makes possible an accession of truthful knowledge and intercommunication between the Physical, Logical, and Metaphysical, and from this the accession of truthful knowledge augments the performance of altruism. Altruism in turn enhances our capacity to know and to evolve our combined cognitive abilities. We can now 'shine' and simultaneously 'madly' *together* evolve new ideas while respectfully discussing fairies and Gods in a combined infinite progressive morality…and trust the water to be drinkable and to be shared. We are grateful now to exist in an inescapable evolution where our cognition and morality can simultaneously expand as part of a principled network evolving in complexity, being forbearing toward all spheres but functioning confidently in the Physical sphere.

## 5.2 Culture and Art

### *Culture*

Culture and religion have traditionally been seen as completely separate social practices, although closely related. Complex discussions aside, it can be securely claimed that the predominant difference lies in religion having a metaphysical core belief and culture a self-conscious need. Culture has generally been avoided by natural scientists as a topic for social anthropologists and sociologists to take care of. An earlier notable definition of culture came from anthropologist Edward Tylor (1871) who opens his pioneering anthropology text, *Primitive Culture: Researches into the Development of Methodology, Philosophy, Religion, Language, Art and Custom,* with the stipulation that culture is, "that complex whole which includes knowledge, belief, art, law, morals, custom, and any other capabilities and habits acquired by man as a member of society."

Culture was subsequently also proposed to be viewed as 'the total shared, learned behavior of a society or a subgroup.' These dimensions were combined by the late renowned anthropologist Bronislaw Malinowski and first publicized in 1931 in a respected but now equally outdated formulation, where culture is seen as a well-organized unity divided into two fundamental aspects—a body of artifacts and a system of customs.

Although undeveloped for current times, these definitions did, however, open the focus on either external (ecological) or internal (psychological) leanings to follow, exposing it in turn to methodological individualism. We can see, with an array of definitions and with fluctuations between mostly subjective ideas about what culture is, that it can become difficult to define.

Based on our proposal here we can perhaps suggest culture to result from: *perception interacting with an ecosystem stimulating a certain form of social behavior in the Logical sphere*. This is subjective, ecologically influenced and driven by a social need to interconnect and identify (belong) with a group of similar beliefs and ideas. It may act as a safeguard for members of such a group and have some benefits for society at large by at times even stimulating change in sharing new ideas. Simultaneously, it may strongly adhere to beliefs or ideas, creating the potential to clash with conflicting ideas. The social benefit then ceases if it causes violent confrontation and restrains the natural evolution of new and better adjusted ideas.

Culture by the nature of its origins and function is also set on customs and maintained with some element of dogma to retain group order. If, however, a culture is non-progressive and strongly and narrowly defended by its benefactors, or set on unpliable misconceptions, it can adversely affect society and slow down an evolving epistemology. This can result in much conflict and suffering, based on a then maladjusted methodology.

Principally a Logical sphere activity, it affects our Physical sphere with expressions of rituals, setting rules of conduct with the objective of having pragmatic value to act as a 'social glue.' From a culture can develop, besides rules, rituals, and behavior patterns, a normative and ethic. If a culture becomes well established and set on a core belief creating moral values with a large following, it may have a significant impact on society. Although culture is responsible for creating a social outlet with support and the potential to establish interconnections, it can also stimulate segregation. When a culture dominates, set on false and unadjusted values and clashing with other beliefs and ideas, it can cause conflict and much harm to society as a whole. Culture can therefore have a significant impact on social progress when infiltrating science, healthcare, and education.

* * *

In animals, evidence for group-specific innovations, such as group trends in differing nut-cracking techniques (culturalism) among chimpanzees, has been recorded and described by researchers V. Horner and A. Whiten (2005) and in orangutans by C.D. Knott and S. Kahlenberg (2007). This research has indicated that an orangutan can sense a tribal difference between orangutans merely by the way a nut is cracked. Culture and cultural transmission have also been documented in dolphins.

A complex interchange between cultural values and a social group may also exist. A social group may consist of different cultures. Today new cultures are emerging all over under the reigning Internet culture and can easily spread globally assisted by social media and the Internet, with many in search of a sense of belonging. Richard Brodie in *Virus of the Mind* (1996) is renowned for making a comparison of cultural trends spreading in the form of *memes* with some similarity to a genetic spread. Attempts were also made to create a similarity between genetic fitness and cultural fitness where memes (a culture) may either survive or go extinct. This was later open to much criticism, with culture considered by some social scientists as a much more rapid method of *adaptation to evolve social interaction* rather than outcompete others. Such conclusions were made by using methods designed by population biologists and published by Robert Boyd and Peter J. Richerson in the article 'Culture and the evolution of human cooperation' (2009).

After our discussion here, we can now see genes more as propagators of 'ideas' with plasticity in response to the world and social needs, and memes only fitting this picture if such social demands are met and cultures stimulate change and progress. Memes on the other hand, as traditionally associated with, again, that dreaded word *reductionism*, have the potential to reduce social interaction and force uniform behavior and so segregate and isolate groups. Brodie was right in perhaps comparing a meme to a virus of the mind with the potential then to damage the plasticity needed to evolve progressive social structures in an expansive network.

At its most extreme, culturism directed by a metaphysical core belief can affect our existence by means of extremist groups. Religious and cultural wars have inarguably been responsible for more 'unnatural' death and suffering than any other cause. We can also (depressingly) state that, whereas nuclear weapons, unethical genetic-manipulation and SARS epidemics have the potential to objectively cause large-scale mortality and much suffering, cultures and religion (based on tradition or false beliefs) may more subjectively do the same. The ongoing fundamentalist religion-driven threat to society is in no need of more evidence than following the daily news. The problem clearly starts where culture and false beliefs interact with a Logical sphere and infect a healthy Physical sphere.

We can also see how from a culture can follow a set of rules of conduct, such rules setting a normative. From a normative setting moral values, we can create an ethic. So, it would not be wrong to state that wearing expensive branded 'hip-hop' outfits in an affluent part of suburban Sydney is an ethic or 'moral value' that identifies with the struggle of urban 'ghetto' living (albeit well removed from it). Or a suit and briefcase with a Wall Street corporate culture, belonging to a 'monied elite,' or sneakers and expensive branded casual wear with Silicon Valley. We can see how both religion and culture are belief systems with a moral impact that aim at identification with a social group but can also trigger segregation, with the potential then to promote grudges and even conflict between groups. Ideally, we should progress to harmonize cultures, reduce segregation and respect diversity in ideas. Culturism is only beneficial if it assists in establishing principled interconnection and helps spread ideas to meet united concerns and outcomes (in the PSR)—adapting faster to changing social demands. Enforced suppression of a culture, its values, expressions and rituals is certainly *not* what is suggested or called for, as it will merely become guilty of the same crime it is trying to suppress. On the other hand, showing respect for diverse cultures, working together to co-evolve and to mutually *benefit* from the marriage of creating new ideas to circulate in the Physical sphere, can be seen as a moral duty. We can lastly use a rather simple and perhaps more light-hearted example to bring the point home in conclusion on this topic.

If I believe in fairies and advocate their existence based on seeing them in my garden and as a result create fictional entertainment or a culture around this imaginary practice, it may have some harmless social value; think of Tinkerbell. If, however, I start killing innocent people because I believe they hurt Tinkerbell or do not believe in fairies, this has much more serious social implications.

If fairies enter with uncertainty into my Logical sphere, I can perhaps be considered not entirely irrational if not eccentric; if, however, they affect my own or others' Physical spheres, I can be considered mentally unwell.

## *Art*

> *Art, then, is an increase of life, a sort of competition of surprises that stimulates our consciousness and keeps it from becoming somnolent.*
>
> Gaston Bachelard, *The Poetics of Space*

A comprehensive topic in itself, art is closely linked to culture and in many ways reflects the ultimate expression of the emotion of a society and culture. We have ample historic and contemporary examples here. We can all relate to the numerous paintings presenting the harrowing scenes of the Crucifixion as seen in art galleries around the world (the effect of religion on the arts is significant),we have Impressionism, Pointillism, Pop culture, Hippie culture... to mention only a few from a long list.

Certainly, we cannot imagine a world or even necessarily want to exist in a world without art and some diversity. We know how, through the centuries, the arts in all their different forms, be it graphic, literature, music or presenting our aspirations through the performing arts, have been invaluable in our search for meaning and as an outlet for human emotion. Art quite clearly can express and reflect the current state of the human condition. It draws on all spheres of reasoning to represent and relay our endeavors and asperities in such a manner that we can sense both the beauty and acrimony involved in living. In this way, then, art becomes a means to communicate and interconnect and circulate both ideas and emotions between all three spheres of reasoning to liberate our higher cognitive function from regimen. We can see art as also vital to secure pliancy and raise us to a higher perceptive level more open to the metaphysical, while serving as protection against authoritarianism and control, the latter two the worst enemies of a liberal and progressive perception.

We do, however, face another conflict here on closer scrutiny of this relationship between culture and the arts. The first evidence of this conflict originates as mentioned, in that cultures can stagnate new ideas, especially when backed by false beliefs with mass impact. In contrast, art in most cases has the potential to galvanize new ideas. We have seen how harmonizing beauty, truth and goodness can bring us closer to beauty than a singular and superficial approach possibly ever can. We can also sense the attractiveness of aspiring to a belief in a deity of infinite goodness, benevolence, and wisdom, promising us hope and escape from all the strife, unfairness and suffering we witness in the world around us. We can clearly be forgiven for our blind faith and ambitious search for any evidence to back such a benevolent belief, and for art historically so desperately searching for escape here. We do, however, still sense some insecurity when such a benevolent search inevitably still lacks any sufficient evidence. In this, then, we see also the main difference for us here between culture and the arts; with the arts attempting to liberate us from the reductionism that results from fixations with a set object, belief, or matter alone, art bravely opens itself to the metaphysical.

The beauty of art then is also seated in this unlocked imagery, harmonizing goodness, truth and beauty as discussed before, giving us new hope. Culture in contrast promises hope and protection from what we sense is lacking in truth, goodness, and beauty (fear) in an attempt to gain control and establish conformity through segregation. In this attempt to isolate itself from wrongs and injustices also lies its omission. Both have metaphysical elements, but unlike false belief systems or cultures based on such beliefs, art openly denies fixations and questions the truth by expressing truth and beauty as it is *imagined* (in its beauty) by the artist. Art therefore has the potential to bring us closer to unobtainable truths, albeit abstract, more so than a non-progressive culture is able to.

## 5.3 Morality in Biochemistry

> *After having followed the day-dreams of inhabiting these uninhabitable places, I returned to images that, in order for us to live them, require us to become very small, as in nests and shells.*
> Gaston Bachelard, The Poetics of Space

I have previously hinted at a moral order extending to the intra-cellular world and will attempt to clarify this a bit more to avoid being accused of unrealistic eccentricity or guilty of the so much dreaded charge here—reductionism. I draw on the very familiar citric acid cycle as one of many examples that can be used in biology.

Marc Bekoff and Jessica Pierce have defined morality more recently as:

> a suite of interrelated other-regarding behaviors that cultivate and regulate complex interactions within social groups. These behaviors relate to well-being and harm, and norms of right and wrong attached to many of them. Morality is an essentially social phenomenon, arising in the interactions between and among individual animals, and it exists as a tangle of threads that holds together a complicated and shifting tapestry of social relationships. Morality in this way acts as social glue.
> M. Bekoff and J. Pierce, *Wild Justice: The moral lives of animals* (2009, p11)

We can see from this more apt definition by Bekoff and Pierce, when compared to other more historic attempts, how morality can subsume inter-perceptively as follows. Patterns of complex interactions relating to wellbeing or harm can be seen on the cellular level and throughout nature and on the physiological level where cells and their protein and biochemical orchestrators can be seen as members of a 'micro-social structure' constantly communicating with each other. We can say that cells and biochemical cycles (like all life) are dependent on complex principled interactions to create an *ordered outcome* befitting the interactions between *a*'s and *b*'s and have a set of rules ('morals') to adhere to.

Let us take the classical energy-producing citric acid cycle inside a cell, as in Figure 3.

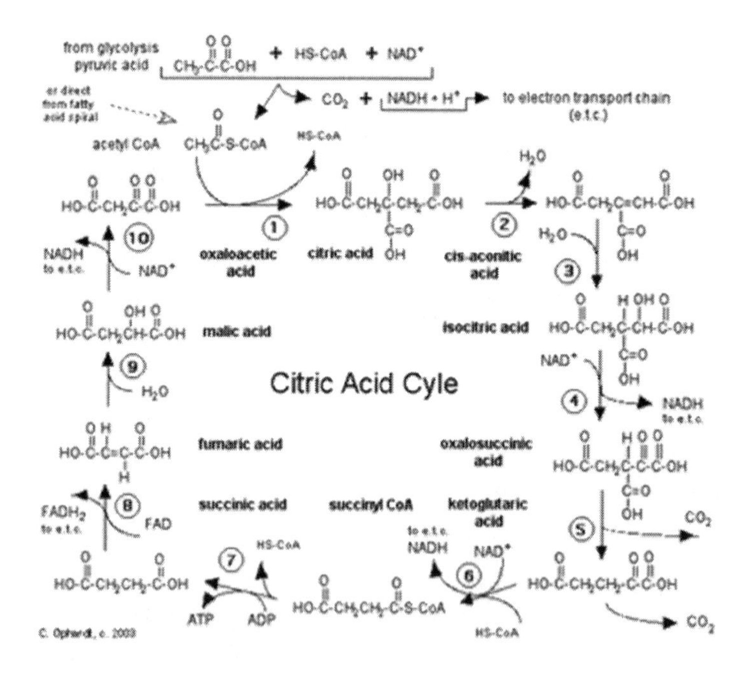


Readers do not have to be troubled with the biochemical detail here or understanding beyond what we can be seen as an ordered, regulated system of harmonized organic chemical interactions. It is notable that specific 'behavior' will result in principled energy production and reprobate interchanges will not. Such organized intra-cellular chemical interactions in turn support organelles, organs, and all organisms.

Should external influences (a new culture) cause aberrant behavior and these, mostly protein molecules, are influenced to behave in a different manner (for instance in consuming the protein L-serine), the ordered structure will act differently and affect the overall outcome of the bio-structure, including the organelles, organs, and eventually the entire organism. This dissonance could result in complete *dysfunction* (no production and cell death), *reduced function* (less production), or less commonly in a function that may be better adjusted to changing environmental demands and seen as *increased efficiency* with *a new idea* (similar to genetic expression in the evolutionary process).With the citric acid cycle used here as only one basic example of many other complex interactive bio-systems we have in common with other organisms, we see a pattern of 'social demand' for the cycle to function efficiently—and yet still have an outlet to be responsive to change. We can also claim, backed by support in my argument here, that Bekoff and Pierce's definition of morality can appertain to the citric acid cycle.

* * *

Beyond the cellular and physiological level and perhaps on more generally allowable ground, many researchers have come to conclude that animals have morality or the ability to engage in moral and immoral behavior (de Waal, 2006; Bekoff & Pierce, 2009). De Waal claims that "empathy and reciprocity are necessary preconditions for morality, these behaviours to be evidence of empathy in chimpanzees" (F. de Waal, *New Scientist* 192(2573), pp60–1 [14 Oct 2006]). Researchers Bekoff & Pierce got straight to the point in the opening chapter of their book *Wild Justice* (2009, p1), and stated "that animals feel empathy for each other, treat one another fairly, cooperate towards common goals, and help each other out of trouble. We argue, in short, that animals have morality." After a lifelong career as a practicing vet, for me what intuitively follows is, how can they not?

We may also argue that an animal who lacks many of the cognitive capacities of adult humans can still be a 'moral agent' because there are different kinds of moral agents, and animal species can therefore have their own form of morality. This view, where even on the physiological level a moral agent (correct action) is required as a necessity for a cell to function in an organ in concord with an organism, should not be seen as extending moral individualism to the cellular level or as part of a moral hierarchy. Neither is it a call for a form of eccentric moral collectivism where cells or molecules set our moral code. It should rather reemphasize the need of *dependable interconnections on all levels to progress a cognition based on principled, perceptive interactions on multiple levels* to achieve best outcomes for all in unremitting change—this ethic extending throughout the phylogenetic tree. We have also discussed that although orderly conduct is required for systemized harmonized interaction, pliancy is also available through mobile DNA when required, as demanded by continuous environmental changes.

Despite obvious differences, others and I here claim more non-individualistically that the important similarities between species (and cellular structures) include the capacities for empathy, altruism, cooperation and perhaps a sense of *fairness on all levels*. Whether or not such claims about animal capacities are true is a matter of much current research but is irrelevant. The irrelevance lies in the fact that the objective evidence of species differences is as obvious as are the biological and physiological similarities. So, if morality stands for respect for others or another's ideas, we cannot accept a morality with differences (as mentioned before in the opening chapters of this text) based on a biased search for behavior disparities and not integrating the similarities. A true morality cannot be discriminatory, regardless of who or what or even species; it has to be interconnected, progressive and all-regarding. This logic can further support the call for not subjecting any form of life to preventable or unnecessary suffering, and this can be further extended to not harming a shared environment. We now also understand that, inescapably, it is pointless to attempt to isolate ourselves from such an emerging noxious environment.

A pragmatic universal ethic will have to be equally diligent in preventing a laboratory rat's suffering under human-inflicted experimentation or a homeless person dying in misery without care or assistance on a street corner, or else fail on all levels. Neither can it fail to have life potentially benefit or be saved by testing a life-saving anti-serum against Ebola on a rat under *humane* conditions or neglect to consider the environment in our actions. Likewise, such a morality will not make a cell routinely self-destruct or let it be indefinitely influenced by manipulation or a harmful culture without evolving an escape from such misconduct. I proposed here how such moral decision-making can be practiced with reduced complexity while avoiding reductionism, using perceptive yet pliant interchanges between Physical, Logical, and Metaphysical spheres of existence, operating sensibly, pliantly, and freely under a judicious evolution as:

$$Ev\ (mo)=\sum\infty\Delta C\{\infty\Delta a(\text{Metaphysical}\rightleftharpoons \text{LSR}\rightleftharpoons \text{PSR}) \approx \infty\Delta b(\text{Metaphysical}\rightleftharpoons \text{LSR}\rightleftharpoons \text{PSR})$$

It stands clear now that a principled universal interconnectivity traced down to the cellular and atomic level was responsible for evolving group social, emotional and cognitive progress. And the health of this network depends heavily on empathy, altruism and fairness on *all* levels, stretching to the environment. With no excuses now, and no need to further here discuss a rather dated Kantian 'human rationality' as lacking in animals, we can also move beyond the basic utilitarian idea of 'kin altruism' and 'reciprocal altruism,' as a genetic trend only determined by natural selection. The inadequacy of kin and reciprocal altruism seen as merely useful behavior trends, genetically favored and established through natural selection to explain its prevalence, can now be seen as only part of a much larger motive. We can accept such concepts as having pragmatic evolutionary value to small groups, but more liberatingly now also see it as a universal requirement of a perceptive evolution. If we take a simple utilitarian stance on morality, where moral conduct is perceived as promoting the general good or welfare of the community to improve its potential to survive, we may also see this as an insecure way to define kin and reciprocal altruism beyond reward of kin and reciprocal reward, with obvious limitations. We also have to define, or perhaps redefine, community today; clearly again, all this stands for something much more comprehensive.

Community today, set in a 'global village,' is unlike what it was in preceding eras and is changing within a rapidly evolving epistemology all the time. If altruism and moral behavior are seen as set dogma within a hierarchy with utilitarian value to an elect group, we face serious problems in a new era where we all confront the same global issues with the same genetic roots—with ultimately nowhere to hide but reliable and progressive knowledge in our PSR. Should we attempt to define an individually-set morality for different communities, subgroups, individuals and situations left open to be argued and swayed by authorities and elect decision-makers, we cannot operate such discriminatory systems practically or fairly. This in turn will secure a morality in dogma and, besides the complexities, already set in equivocality, it will create ongoing conflict in a schema we also cannot define as morality. However much we try to sway and defend our definition of what morality is or what we think it should stand for, it remains a concept much wiser than us and driven by a progressive immeasurable perception.

Without much arguing necessary, we can clearly see that both altruism and morality also stand for something much more universal and larger than can be set by defining genomes, community or culture and their individualized needs, and that there can be no such thing as moral inequality or moral externalism. And yet we can now understand how morality co-evolves as an interconnected perceptive network.

Furthermore, what counts as fair may also differ, at times vastly, within such a network—what is unfair to humans may not be unfair to another species. It was suggested by Marc Bekoff and Jessica Pierce in *Wild Justice: The moral lives of animals* (2009) that researchers should consider natural behavior in order to uncover potential fairness norms.

This was also reiterated by Sarah F. Brosnan, L. Salwiczek, and

R. Bshary in the article 'The interplay of cognition and cooperation' (2010).

* * *

The mechanisms of altruism, cooperation and punishment, the existence of social norms, the affective requirements of moral reasoning—all require interactive perceptive means as well as empirical investigation of prosocial behavior in demand of a progressive ethic. Although much research is still needed, it can be clearly seen that even the possibility of morality objectively analyzed in a stratified or isolated manner in science, or anywhere else, now becomes almost impossible without recognition of the interconnections in a perceptive evolution. This in turn, then, is also in the context of a pragmatic Physical sphere variable with time, species, individual, and place, and reliable interchanges with the Logical and Metaphysical spheres.

There will be contemporary arguments against such a proposal, some even from scientists prominent in their fields such as psychologist Helen Guldberg (2011), in extreme claiming that only humans have morality and not animals. "Human beings, unlike other animals, are able to reflect on and make judgements about our own and others' actions, and as a result we are able to make considered moral choices. We are not born with this ability," Guldberg argues. We can perhaps also agree that we are not born with it, *but* confidently claim that we co-evolve it transgenerationally as an interconnected, progressive, perceptive, living network—operating both internally and externally.

Other opponents will have their roots in creationism perhaps, where belief at times overshadows logic. Either way and in respect of all personal opinions, I fail to see how we can, in any current-day objective science and with new understanding, arrive at any conclusion on morality if reverting back to metaphysical beliefs, pseudoscience, external moralism or moral individualism while disregarding our evolutionary roots. This has now with new knowledge become exceedingly difficult in an age of interconnection where reductionism is becoming more and more insecure as our only future base, as now evident in both the physical and natural sciences.

We have stated how social cognition and the need for interconnection can be traced to the atomic, physiological and cellular level in our Physical sphere. On the subatomic level in particle physics we have become quite familiar with quarks and leptons and their interplay in weak, strong and electromagnetic forces. We can even be forgiven for saying that these particles are 'aware' of each other, constantly interacting and dependent on each other and their environment for overall outcomes. We can securely conclude that even on the particle level behavior is determined by the presence of other particles and physical forces. Physicists are beginning to see how, in now nearing the limits of a reductionist approach in measuring these forces and particles, we will continue to have much unanswered. All we may end up knowing with some level of certainty is that demand is placed on an interconnected web to perceive change and behave in a certain pliable manner while it engenders its own complexity. Be it a biochemical cycle or a social group, we have the establishment and vital need for morals to drive a perspicacious evolution. With such an approach we can allow for some level of progressive moral particularism, but not accommodate moral externalism or solipsism. Our moral duty is to grow our wisdom by progressing our morality and understanding and *not* setting material barriers to such a universal expansion.

Based on this, if evolutionary scientists state that natural selection has favored morality and altruism, they are not completely wrong. If they say, however, that a progressive morality and altruism set the basis for evolution to function as a living network, I believe they are on target. We can also state that there is an 'evolutionary ethic' even on the particle level that extends up the phylogenetic tree to us humans, requiring us to behave in a certain harmonious interconnected way with some pliancy to continuously evolve our perception, aimed at an overall more benevolent outcome for all. This demand subsumes to cellular biology, physiology and physical particles as discussed.

Definitions of society and community itself may and will continuously change with time but, as our cognitive abilities evolve, I believe so will our morality advance, as we have *a priori* seen since Darwin's return from his voyage on the *Beagle*. Darwin, a kind-hearted person, strongly anti-slavery and firmly against 'uncontrolled' vivisection, wrote the following in the *Gardener's Chronicle* in 1863:

> If we attempt to realize the suffering of a cat or other animal when caught, we must fancy what it would be like to have a limb crushed during a whole long night, between the iron teeth of a trap, and with the agony increased by constant attempts to escape.

Noteworthy and in contrast is Nietzsche's unrealistic argument against establishing a general morality for all people: "the question is always who he is, and who the other person is..." (Nietzsche, *Beyond Good and Evil*, Aphorism 221). And how, we should earnestly ask, can moral behavior be based on anything other than selfless concern for others? He confidently continues:

> Every unegoistic morality that takes itself for unconditional and addresses itself to all does not only sin against taste: it is a provocation to sins of omission, one more seduction under the mask of philanthropy—and precisely a seduction and injury for the higher, rarer, privileged.

Nietzsche's disparate attempt here to justify inequality and consider a normative not applicable in the same manner to all persons, and we can now add all lifeforms, disqualifies it from having any useful contribution in a realistic search of morality, or any ethic for that matter. His radical approach, as historically suggested, was perhaps more driven by a need to please a contemporary elite and their preferences. Set in an underdeveloped science, rather than an authentic universal concern, his attempt to build subjective phrases like "sin against taste" into morality is perhaps ludicrous besides being unrealistic. Creating Nietzsche's proposed, and ill-defined, 'superman' will also subjugate the demand for such a 'superman' to operate under a 'super morality.'

We may finally, on this unrealistic approach of a Nietzschean morality, claim he argued that pyruvic acid is more worthy of a place and of higher importance in the citric acid cycle than oxalic acid. It is doubtful that a Nietzschean-based citric acid cycle would produce a smidgen of utilizable energy, malfunctioning as an individualized 'super-morality' of higher importance than the holism of a citric acid cycle or body complete.

Perspicuously, it can be seen how an adaptable unanimous evolutionary cognition and morality operating on *all* levels is also then well equipped to progress beyond the hampering effects of false belief systems, intellectual dogma or suppression, as it is based on interaction between all spheres of reasoning.

In conclusion of this chapter we should ask ourselves: Do we want to continue to have conflicting cultures, religions and shaky belief systems fight it out to see if a realistic normative can survive such confusion or perhaps be secured by a Nietzschean superman, or more likely superpower or mega-corporation? Or should we perhaps accept that, at the very best, what we can all hope for with such a system is to violently replace one false belief with another, creating needless and endless suffering and delays in our cognitive and moral progress. We can see the urgent need for science (including a non-fraudulent, pragmatic medical science) under a new ethos to join hands and discuss this important issue on all levels much more progressively. Education, practice, and research all need to readjust their ethic as pivoting around the current economic structures to better understand and adapt their future paths in such a cognitive network. After all, we are pliant and temporary idea-makers in a progressive cognition, and the more honest and higher the morality that these ideas originate from, the more beautiful the world will be for all of us, our children, and grandchildren.

# 6

# Science and Critical Thinking in Interactive Spheres of Perception

Now, as part of what can be seen as an almost 'rhapsodical' evolution with its receptiveness, pliability and adaptability, we have the promise of a sagacious and sensitive cognitive destiny. It has become increasingly difficult, if not impossible, to imagine an evolution that is non-perceptive or not consecrated to embalm a universal genetic code secured against sublunary corruption, moral decay and vanity. In this security remain our hope and progress.

We simply cannot, as seen throughout this book, evolve authoritative and dependable knowledge based on science alone and with belief as our standby elixir under a maladjusted economic structure. An epistemology ignoring the evolutionary roots of our perception is even more equivocal, insecure, and procrastinating. There exist vast and meaningful questions outside the confines of empirical science begging for advancement in an ethical arena where realistic knowledge can candidly revolve malleable and functional ideas between *all* spheres of perception. With the uncapturable truth now circulating *between* these changing particles, theories, and material world we so desperately try to capture and reduce in complexity, a pragmatic and universal morality can be grown. This can only happen if we liberate and evolve trustable ideas between interactive progressive spheres of perception, in accord with an evolving morality, interconnected and free of material fixations and unbendable theories.

Conclusions are more likely to be conducive to produce results remaining faithful to the overall improvement of the human condition when the ongoing conflict in our Logical sphere evolves from a sober ethic and our Physical sphere operates in a science free of fraud and manipulation. Not only is the future of science and healthcare dependent on perspicaciously instituted foundations supported by a truthful ethic to secure the reliable interchange of knowledge, but also life as an interconnected progressive perceptive force. Our challenge is, and will become increasingly so, to introduce through our educational systems and public platforms a truthful, all-inclusive pliant knowledge-base that can be trusted and accepted on all levels and across all cultures, religions, and socioeconomic groups striving for a universal morality. Such a responsibility will not be penurious and vulnerable to dogma, restrictive cultures or false belief systems; it will axiomatically flow from an unanimously evolving cognition with better understanding of how things work in an interconnected perceptive drive. A greed- and fear-driven society will have to be replaced by a conscious one with pragmatic and dependable knowledge and understanding of our existence as part of an integrated universal whole, operating under a sagacious ethic.

Open critical thinking has mistakenly in recent years been piqued by a materialistic society as failing to have any applicable and practical value, perhaps even the potential for radicalization or to threaten powerful entities. I think our educational facilities and society in general have never been more in need of pragmatic attentive free-thinking, to not only help reduce bias in guiding a more progressive knowledge, but also to urgently reduce the emergence of a pseudoscience supported by blind profits. Under conditions of fairness and with evolution now unraveled as a perceptive and principled drive, science backed by critical thinking with unimpeded inhibitory control can now help to evolve a progressively more interconnected global society. In such an advancing epistemology also constantly evolving new fairness norms, we can all prosper and benefit from emerging knowledge and technology, rather than be driven as 'desperate survivors' relying on control of the material world or metaphysical beliefs as our saviors.

There are a good few solid reasons why we should urgently kindle progressive critical thinking at our teaching facilities free of trends, culturism, false beliefs and greed in all spheres of perception. I list only a few as they came to mind:

1. Society has inevitably reached the point of being intermingled in a global community. This global interconnection was initially mainly due to advances in air travel and was industrial based, but in the last few decades it has been enhanced by massive advances in technology—the Internet, mobile phones, television, cinema, an internationally transferable workforce, circulating music, and culture, to name but a few. More recently, with new scientific advances and better understanding of our vast universe, we are also starting to see our existence as universal, with Earth as our only home for now.
2. As philosopher Karl Popper and many others have suggested, the empirical sciences are systems of theories where these theories can be seen as 'nets cast to catch what we call the world.' We can now furthermore see how, in an expanding universe and with a perceptive evolution, all our attempts at making the mesh finer and finer is never-ending, because the net is forever stretching.
3. People have an innate need for an ethical, just, fair and equal society in order to live happy, meaningful, and fulfilled lives, and once born they have the right to such basic needs. With different cultures now more freely interacting, the clashing ideologies and dogma expressed within and between different religions can only be seen as segregationist, egocentric, and aimed at manipulating power and control to disrupt the evolution and expression of more progressive naturally occurring social demands (and moral evolution). The need for critical thinking on these matters is more urgent with everything now refreshingly seen (regardless of position or status) as parts of an interconnected network, as we emerge into a post Darwinian less materialistic era.
4. Fissiparous religions and cultures are inarguably sluggish to adjust to such needs and are losing support because they cannot conform without radically changing their innate core beliefs. Furthermore, it is becoming more questionable how one

on their innate drive to control. Should this perhaps change they would not tolerate inequality, and in turn would have to surrender control to a concordant and universal morality and ethic as firmly argued here.

8. Critical thinking to open all spheres of perception is one potential method to escape from our current reductionist approach centered around profitability and its animus, with more potential to help reduce bias and corruption.
9. Interconnected spheres of perception, in the context of the explanation proposed here, represent an open avenue to an evolving new epistemology allowing flexibility yet remaining secured in reliable and pragmatic knowledge. This serves science well as its more fixed objective security in positivist empirical knowledge is becoming less meaningful in light of its (science's) own discoveries.
10. A pliant ethic, with critical guidance and support gained in a truthful science, will have more credibility, with the potential to be introduced into commercial fields and politics where ethics and morals have become lacking to a point of 'ethical' being taken as far as swayable by costly legal services and what you can get away with.
11. Any ethic with a realistic, pliant, and pragmatic base and universal application will potentially save more lives and prevent more human and animal suffering than any new profit-focused medicine or medical technique in the near future can potentially offer. Sadly, *hardly any funding and little effort is directed at promoting such goal-directed research.*
12. What other options do we have?

The challenges facing our current and future post-technocratic society can certainly be seen as a big task with material limitations when we consider some of the ethical implications, but not as idealistic or despondent if seen as part of an integrated evolutionary morality and cognition—interconnected in three principled spheres of perception.

# References

It is impossible to recall or hail all the wonderful productions over many years that stimulated the concept to emerge as presented in this book. Cited or consulted works during this composition are indexed below.

Althusser, Louis. *For Marx.* London, Verso Classics, 1996.

Aristotle. *Nicomachean Ethics.* Cambridge, Cambridge University Press, 2000.

Armstrong, J. *The Secret Power of Beauty.* Penguin Books, 2005.

Armstrong, J. *Love, Life, Goethe.* Penguin Books, 2006.

Ashton, J.F. *In Six Days: Why 50 scientists choose to believe in creation.* New Holland, 1999.

Bachelard, Gaston. *The Poetics of Space.* Ed. 1994 with foreword by John R. Stilgoe. Originally published: New York, Orion Press, 1964.

Barkow, J.H., Cosmides, L., Tooby, J. *The Adapted Mind: Evolutionary psychology and the generation of culture.* New York, Oxford University Press, 1992.

Bekoff, M., Pierce, J. *Wild Justice: The moral lives of animals.* University of Chicago Press, 2009, pp1-6.

Berge, H.M., Gjelstad, S., Furu, K., Straand, J. 'Use of glucosamine does not reduce the need for other pain-relieving drugs.' *Tidsskr Nor Laegeforen,* 130(15), pp1463–6. 12 August 2010.

Bergson, H. *Creative Evolution.* Dover Books, 1998.

Bernier, P.J., Bedard, A., Vinet, J., Levesque, M., Parent, A. 'Newly generated neurons in the amygdala and adjoining cortex of adult primates.' *Proceedings of the National Academy of Sciences of the USA* 99(17), pp11464–9. 20 August 2002.

Black, C., Henderson, R., et al. 'The clinical effectiveness of glucosamine and chondroitin supplements in slowing or arresting progression of osteoarthritis of the knee: a systematic review and economic

Perhaps the post-reductionist era, confronting limitations in the current ambitious pursuit of technology, will evolve us into a more perceptive moral consciousness, as the hallmark of an enlightened new era full of potential.

7

# Conclusion

*The darkest places in hell are reserved for those who keep neutrality in times of moral crisis.*
Dante Alighieri

*The philosophers have only interpreted the world, in various ways. The point is, however, to change it.*
Karl Marx

I hope it is now evident that in our emergent society we need to replace an outdated disjointed concept of evolution with a more interconnected perceptive and moral one. In harmonizing and circulating our perception between reliable spheres of perception as discussed in this book, and by acknowledging our subservience to a progressive universal ethic, I hope I have presented to the reader a more balanced outlook on life and better understanding of our evolutionary construct.

We can now see evidence of approaching the next stage of our evolution as more consciously aware of the interconnectivity and universality required to operate within a perceptive network. In this post-reductionist era with progressive knowledge undressing us as subject to an interconnected and principled perceptive network, our moral demands have also escalated to a new level. The urgency of circulating trustworthy and pliable knowledge between all three spheres (Logical, Physical, and Metaphysical) is also where we can proximate morality as a universal duty—with our knowledge as sound as the dependability of these interconnections.

Now preserved from mutability, a unanimous global perceptive drive with an unrestricted intrinsic evolving morality can face change as a *united* concern, and benevolence and altruism can emerge and evolve a post-Darwinian era into a post-reductionist one of progressive religion can be more 'right or wrong' than another and how conflict between religions can be called religion. The perspicacity and truthfulness of science as our talisman in turn is also severely challenged by the aggressive marketing and financial and political manipulation of this more pragmatic attempt in our search for a better and rational world.

5. Any moral code, ethic or idea that can 'glue' a global community together in fairness and equality will have to rid itself of dogma and historical links to suppression, corruption, greed, bloodshed, or links to suffering because of earlier actions. It also has to continuously evolve itself. Both religion, ruthless economic strategies and politics have inarguably lost their credibility in persistently associating themselves with such despotism. Even science has been linked to bloodshed and fraud, so this leaves an 'open' new knowledge exposed to critical thinking in a sound ethic as our only realistic defense—which is why *Spheres of Perception* was born.
6. We need some protection and guidance as an advancing technocratic society where an invaluable technology is slowly becoming more valued than a human workforce. The pragmatic value of such technology is in turn also mainly motivated and subsequently threatened by marketability and profits, more than how it can benefit humanity overall. The ethic there has to change.
7. It seems to be increasingly impossible to marry any ethic or moral society together with a system where a few major beneficiaries reap the profits from the effort of the masses in uncertain times. This is when the major benefactors do less for more, with the power to trickle down profits and control the masses (here it seems we also never learn from history). Furthermore, it is impossible for the ethics and moral standards of the major beneficiaries to be accurately assessed, or ever be proved to be superior to those of the disadvantaged group, which again, as is shown by history, they mostly are not, based

consciousness.

Understanding life as an interconnected web, and reliant on a healthy perception and moral progress, we can without discrimination or hesitation start nurturing our own cognitive health and that of other sentient beings *and* the environment, so we can all justly share in its infinite beauty. Disease will be interpreted as anything that may inflict damage on these interconnections in a perceptive network, with its health as strong as its weakest links. Such recognition will extend into a fight against inequality, greed, war, poverty, and fraudulent behavior on all levels of our society as we now understand it as a vital need for a perceptive evolution to continue.

Training and teaching in a critical understanding of a healthy and ethical flow of perceptions between principled spheres of reasoning with pliancy, tolerance, and respect can help set us on the road to an amazing new era of knowledge expansion—reaching into the abolishment of poverty, corruption, fraud, and war. All healthcare workers, scientists, technologists, philosophers, critical thinkers, environmentalists, lawmakers, teachers, and workers on diverse topics and levels will interconnect with renewed focus on the universality of things, in a comprehensive non-greed-driven epistemology—with Artificial Intelligence there to assist and not threaten us. Any damage to this delicate network and its cognition will be seen as a significant moral crime, extending to our environment and future generations and punishable by a *moral* law.

Future research would be more focused on sensory and cognitive wellbeing of a *universal* network, and teaching seen as progressing knowledge on how to maintain harmony and conduct ethical interchanges between all spheres of perception of our existence. A progressive new global society will understand and see itself as part of an interconnected complex, instead of a primeval system narrowly focused on poorly defined profits and boundaries with survival based on equivocal self-centered goals.

Vast new employment opportunities can open up with local communities now not subject to larger enterprises for their livelihoods. A new global workforce with regained respect will have much more time for caring for the needs of all, including the frail, elderly, the physical and mentally disabled, and the environment we so much exploit and isolate for profit. Tapping all the benefits of an advancing technology available, work will be individualized to create time for such social interaction. Such spare time can be more pragmatically spent growing and nurturing edible and organic gardens to supply personal food needs with focus on reducing waste as an interconnected concern with our mother planet, rather than concentrating profits and labor to sustain mega-corporations. Greedy enterprises, bullies and technocrats, profiting from abstract and manipulated regulations to mass-control resources in notional markets, will be lowered in status and penalized to a point where they will have to surrender to more ethical conduct. The ability to cover up the environmental impact of their actions due to financial and political power will also not be left unpunished.

Operating under such a new ethic, all workers will be part of a receptive wealth with education focused not only on knowledge, but on its ethical distribution and application. In acceptance of our interconnection in a united drive where there is little scope for greed, fraudulent conduct or inflated egos, those in need of support will not be left behind. Caring and altruism would now overshadow egocentric behavior, or any self-centered actions driven to gain power and wealth in a society that profit or favor those who help themselves at the cost of others and the environment. Our evolving epistemology would be understood and recognized as part of an interconnected perceptive drive and acknowledged as a fragile but invaluable link in an infinite cognitive progression.

The above scenario undoubtedly will not escape being accused of extreme idealism by some—mainly those finding sufficient comfort in the short-term security that materialism may offer. However, this can now be shielded from such an accusation by being a *goal-directed exigency with a united aim*. In unison with a universal ethic, we can no longer afford to see this as idealistic, but as vital for our continuation as perceptive beings and for the progression of workable knowledge.

The other less attractive option is to confront an outcome as predicted by Karl Marx—with a "history that first repeats itself as a tragedy then as a farce with too many useful things resulting in too many useless people."

I believe the argument presented here, backed by a growing volume of new evidence, can now no longer be skipped over as being idealistic or unrealistic. Neither is there much time for accusations of idealism to be tolerated, as revealed by other avenues explored by science, such as climatology and healthcare. I have aspired to create objective values for practical application and simultaneously to steer clear from reductionism and inflexibility to stimulate ongoing change. Entering an era now with an evolution understood as a more adaptable, respectful and principled system, we have a progressive perception able to confront climate change, unemployment, and other urgent issues we can no longer put on hold. Set in a trustable and universal ethic, our future seems more expectant, if not bright.

There are no ends, only infinite change and new beginnings in evolving reliable spheres of perception…

# Note to the Reader

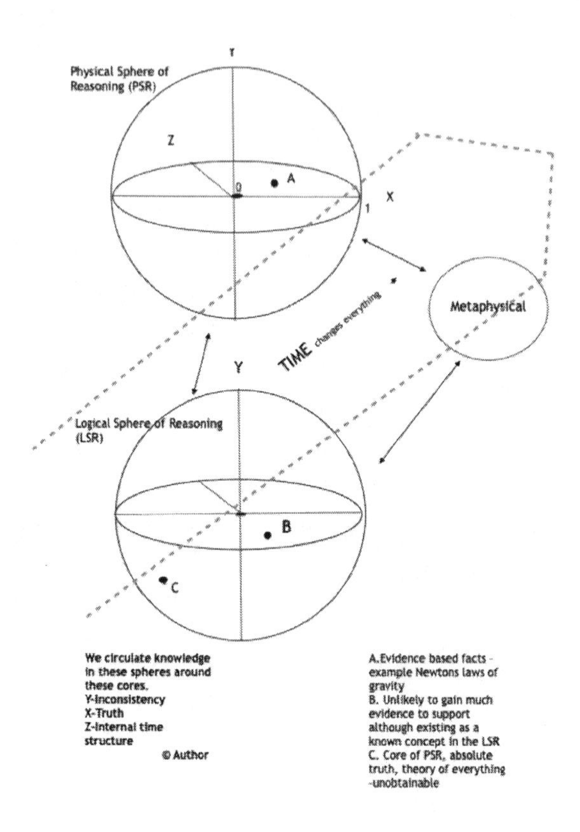


Thank you for purchasing *Spheres of Perception*. My sincere hope is that you learned as much from reading this book as I have in compiling it, and can now share ambitions in a new method to make the world a better place for all. Feel free to add your review of the book at your favorite online site for feedback. Also, if you would like to get in touch and help to progress a growing network of new ideas to advance an improved and united world, please visit my website for news on upcoming works, recent blog posts, or to become a member: https://www.spheresofperception.com/

Sincerely
Theodore Holtzhausen

evaluation.' *Health Technology Assessment* 13(52), pp1–148. November 2009.

Blackburn, R. *Marx and Lincoln: An unfinished revolution.* https://libcom.org/files/robin-blackburn-an-unfinished-revolution-karl-marx-and-abraham-lincoln.pdf (cited Jan 2018).

Blackburn, S. *Truth.* Penguin Books, 2006.

Boag, Z. '*The end of progress.*' *New Philosopher*, Issue 6, pp19, 79–84. Nov 2014.

Boyd, R., Richerson, P.J. 'Culture and the evolution of human cooperation.' *Philosophical Transactions of the Royal Society B* 364(1533), pp3281–8. 12 November 2009. doi: 10.1098/rstb.2009.0134 PMCID: PMC2781880 (cited Sept 2018).

Brain, C.K. 'Raymond Dart and our African origins.' In Garwin, L., Lincoln, T. (eds), *A Century of Nature: Twenty-one discoveries that changed science and the world.* Chicago, University of Chicago Press, 2003.

Brand, E.J., Kennedy, J.L., Müller, D.J. 'Pharmacogenetics of antipsychotics.' *Canadian Journal of Psychiatry* 59(2), pp76–88. February 2014. PMID 24881126.

Brodie, R. *Virus of the Mind: The new science of the meme.* Hay House USA, 2nd Edition, 2009.

Brosnan, S.F., Salwiczek, S., Bshary, R. 'The interplay of cognition and cooperation.' *Philosophical Transactions of the Royal Society B* 365(1553), pp2699–710. 12 September 2010. doi:10.1098/rstb.2010.0154. http://rstb.royalsocietypublishing.org/ (cited 17 July 2018).

Brown, C., Graves T.G., Tucker, A. 'Hyperadrenocorticism: treating dogs.' *Compendium of Continuing Education* 29(3). 2007.

Browne, K.E., Milgram, B.L. *Ethics and Morality: Anthropological approaches*. Society for Economic Anthropology Monographs. Altamira Press, 2009.

Bunnin, N., Tsui-James, E.P. *The Blackwell Companion to Philosophy*. Blackwell, 2003.

Burns, D. *Feeling Good.* William Morrow and Co., 1980.

Campbell, D.T. 'Natural selection as an epistemological model.' In Raoul Naroll and Ronald Cohen (eds), *A Handbook of Method in Cultural Anthropology*, pp51–85. New York, National History Press, 1970.

Campbell, D.T. 'Evolutionary epistemology.' In P.A. Schlipp (ed.), *The Philosophy of Karl Popper*, pp413–63. Open Court, 1974.

Campbell, D.T. 'Evolutionary epistemology.' In G. Radnitzky and W.W. Bartley (eds), *Evolutionary Epistemology, Rationality, and the Sociology of Knowledge*, pp47–89. Open Court, 1987.

Casey, B.J., Durston, S., Fossella, J.A. 'Evidence of a mechanistic model of cognitive control.' *Clinical Neuro Research* 1, pp267–82. 2001.

Cathcart, T., Klein, D. *Heidegger and a Hippo Walk through Those Pearly Gates*. Viking, 2009.

Charlton, B.G. 'Injustice, inequality and evolutionary psychology.' *Journal of Health Psychology* 2, pp413–25. 1997.

Close, F. *Particle Physics: A very short introduction.* Oxford University Press, 2004.

Colcombe, S.J., Kramer, A.F., McAuley, E., Erickson, K.I., Scalf, P. 'Neurocognitive aging and cardiovascular fitness: recent findings and future directions.' *Journal of Molecular Neuroscience* 24(1), pp9–14. 2004.

Confucius. *The Analects.* Introduced by John Baldock. London, Arcturus, 2010.

Damasio, A. *Descartes' Error: Emotion, reason, and the human brain.* New York, Quill HarperCollins, 2000.

Damasio, A. *Looking for Spinoza.* Harcourt, 2003.

Darwin, C. *The Origin of Species by Means of Natural Selection.* John Murray, 1859.

Darwin, C. *The Autobiography of Charles Darwin.* Ed. N. Barlow. London, Collins, 1958.

Davis, A., Robson, J. 'The dangers of NSAIDs: look both ways.' *British Journal of General Practice* 66(645), pp172–3. April 2016.

Dawkins, Richard. *The Selfish Gene.* Oxford, Oxford University Press,

1989.

Dawkins, Richard. *Climbing Mount Improbable.* Norton, 1996.

Dawkins, Richard. *The God Delusion.* Bantam Press, 2006.

De Waal, Frans. 'The animal roots of human morality.' *New Scientist* 192(2573), pp60–61. 14 October 2006.

Dean, K., Joseph, J., Roberts, J.M., Wright, C. *Realism, Philosophy and Social Science.* Palgrave Macmillan, 2006.

Deleuze, G. *Foucault.* Minneapolis, University of Minnesota Press, 1988.

Deleuze, G., Guattari, F. *A Thousand Plateaus: Capitalism and schizophrenia.* Minneapolis, University of Minnesota Press, 1987.

Earman, John. *Bayes Or Bust? A critical examination of Bayesian Confirmation Theory.* Cambridge, MA, MIT Press, 1992.

Eiser, Arnold R. *The Ethos of Medicine in Postmodern America.* Lexington Books, 2014.

Ekman, P. (University of California). 'The argument and evidence about universals in facial expressions of emotion.' In H. Wagner and A. Manstead (eds), *Handbook of Social Psychophysiology*, pp143–64, at p151. Chichester, John Wiley & Sons, 1989. Available at: www.paulekman.com

Ekman, P., Davidson, R.J., Friesen, W.V. 'The Duchenne smile: emotional expression and brain physiology. II.' Journal of Personality and Social Psychology 58(2), pp342–53. 1990.

Ettinger, S.A., Feldman, E.C. *Textbook of Veterinary Internal Medicine*. 6th Edition. Canada, Saunders, an imprint of Elsevier Inc., 2005.

Ettinger, S.A., Feldman, E.C. *Textbook of Veterinary Internal Medicine.* 7th Edition. Canada, Saunders, an imprint of Elsevier Inc., 2010.

*Evangelii Gaudium* (Pope Francis). http://w2.vatican.va/content/francesco/en/apost_exhortations/documents/papa-francesco_esortazione-ap_20131124_evangelii-gaudium.html (cited Nov 2018).

Flavell, John H., Miller, Patricia H., Miller, Scott A. *Cognitive Development.* Prentice-Hall, 1993.

Koestler, A. *The Sleepwalkers*. Penguin Group, 1986.

Kosch, M. *Freedom and Reason in Kant, Schelling and Kierkegaard.* Oxford University Press, 2006.

Kramer, A.F., Bherer, L., Colcombe, S.J., Dong, W., Greenough, W.T. 'Environmental influences on cognitive and brain plasticity during aging.' Review article. *Journal of Gerontology: Medical Sciences* 59A(9), pp940–57. 2004.

Kuiper, M.T.R., Sabourin, J.R., Lambowitz, A.M. 'Identification of the reverse transcriptase encoded by the Mauriceville and Varkud mitochondrial plasmids of Neurospora.' *Journal of Biological Biochemistry and Molecular Biology* 265(12), pp6936–43. 25 April 1990.

Leslie, A.M., Friedman, O., German, T.P. 'Core mechanisms in "theory of mind."' *Trends in Cognitive Sciences* 8(12). December 2004.

Lewis, C.I. *The Ground and Nature of Right.* Columbia University Press, 1955.

Lewis, C.I. *Values and Imperatives: Studies in ethics*. Ed. John Lange. Stanford, Stanford University Press, 1969.

Locke, J. *An Essay Concerning Human Understanding*. Ed. Alexander Campbell Fraser. 2 vols. Oxford, Clarendon Press, 1894.

Lorenz, K. Die Rückseite des Spiegels. Munich, Piper, 1973. [English: Lorenz, K.Z. 1977. *Behind the mirror: A search for a natural history of human knowledge*. London, Methuen, 1977]

Malinowski, B. 'Culture.' In E.R.A. Seligman (ed.), Encyclopedia of the Social Sciences, Vol. 4, pp621–46. New York, Macmillan, 1931.

Malinowski, Bronislaw. *A Scientific Theory of Culture and Other Essays*. New York, Oxford University Press, 1960.

Marx, Karl. *Das Kapital*, 1867.

Marx, Karl. *Capital: A critique of a political economy.* https://www.marxists.org/archive/marx/works/download/pdf/Capital-Volume-I.pdf (cited March 2018).

Matzinger, P. 'The danger model: a renewed sense of self.' *Science* 296(5566), pp301–5. 2002.

McGinn, Colin. 'The Truth about Truth.' In Richard Schantz (ed.), *What Is Truth?* Berlin, Walter de Gruyter, pp194–204. 2002.

Fontanarosa, P.B., Lundberg, G.D. 'Alternative medicine meets science.' *JAMA* 280, pp1618–19. 1998.

Foucault, Michel. *The Birth of the Clinic.* New York, Vintage/Random House, 1994.

Fromm, Erich. *Beyond the Chains of Illusion.* Sphere Books, 1980.

Gelb, M.J. *Think Like Da Vinci.* HarperCollins, 2009.

Gladwell, M. *Blink!* London, Allen Lane, 2005.

Gold, J. 'Cartesian dualism and the current crisis in medicine—a plea for a philosophical approach: discussion paper.' Journal of the Royal Society of Medicine 78, pp663–6. 1985. (PMC free article; PubMed)

Gould, Stephen J. *The Hen's Teeth and the Horse's Toes*. Norton, 1983.

Grayling, A.C. *What Is Good? The search for the best way to live.* Weidenfeld and Nicholson, 2003.

Griffiths M.R., Lucas, J.R. *Ethical Economics*. Third Draft, Tuscany, Oxford and Somerset, June 1995. users.ox.ac.uk

Guldberg, H. 'Only humans have morality, not animals: only humans make moral judgements and moral choices.' *Psychology Today*. 18 June 2011. https://www.psychologytoday.com/au/blog/reclaiming-childhood/201106/only-humans-have-morality-not-animals (cited Feb 2018).

Hamilton, W.D. 'The genetical evolution of social behaviour, I & II.' *Journal of Theoretical Biology* 7, pp1–52. 1964.

Harsanyi, Z., Hutton, R. *Genetic Prophecy*. London, Granada, 1983.

Hauser, M. *Moral Minds: How nature designed our universal sense of right and wrong*. Ecco, 2006.

Hawking, S. *A Brief History of Time.* Bantam, 1988.

Hesselmann, F., Graf, V., Schmidt, M., Reinhart, M. 'The visibility of scientific misconduct: a review of the literature on retracted journal articles.' *Current Sociology Review* 65(6), pp814–84. 2017.

Holtzhausen, T.D. 'Can we develop a pragmatic ethic in healthcare without dogma?' *Proceedings and abstracts*. World Congress on Controversies in Veterinary Medicine. Prague, 23–6 October 2014.

Horner, V., Whiten, A. 'Causal knowledge and imitation/emulation switching in chimpanzees (Pan troglodytes) and children (Homo sapiens).' *Animal Cognition* 8(3), pp164–81. July 2005. Epub 11 Nov 2004. PubMed PMID: 15549502.

Huxley, Aldous. *The Perennial Philosophy*. New York, HarperCollins, 1990.

Huxley, J. 'Schizophrenia as a genetic morphism.' *Nature* 204(4955), pp20–1. October 1964.

James, William. The *Principles of Psychology*, Vol. 1. New York, Holt, 1990. https://archive.org/details/theprinciplesofp01jameuoft/page/n6

Järvilehto, L. 'Pragmatic a priori knowledge: a pragmatic approach to the nature and object of what can be known independently of experience.' *Jyväskylä Studies in Education, Psychology and Social Research* 429, p153. 2011.

Jonsen, Albert R. *A Short History of Medical Ethics.* Oxford University Press, 2000.

Kahneman, D., Slovic, P., Tversky, A. *Judgement under Uncertainty: Heuristics and biases*. Cambridge, Cambridge University Press, 1982.

Kant, E. *Critique of Pure Reason*. http://www.iep.utm.edu/k/kantmeta.htm (cited Dec 2017).

Kay, J. *Obliquity*. Profile Books, 2010.

Kazazian, Haig H., Jr. *Mobile DNA: Finding treasure in junk*. FT Press, 2011.

Kent, K.A., Buckholtz, J.W. 'Inside the mind of a psychopath.' *Scientific American Mind and Brain*, pp22–6. Oct/Sept 2010.

Knott, C.D., Kahlenberg, S. 'Orangutans in perspective: forced copulations and female mating resistance.' In S. Bearder, C.J. Campbell et al. (eds), *Primates in Perspective.* Oxford, Oxford University Press, pp290–305. 2007.

Ko, S.H., Kwon, H.S., Yu, J.M., et al. 'Comparison of the efficacy and safety of tramadol/acetaminophen combination therapy and gabapentin in the treatment of painful diabetic neuropathy.' *Diabetic Medicine* 27(9), pp1033–40. September 2010.

McMahan, J. *The Ethics of Killing*. Oxford University Press, 2002.

Merzel, Dennis G. *The Eye that Never Sleeps.* Boston, Shambhala, 1991.

Mettke-Hofmann, C., Gwinner, E. 'Long-term memory for a life on the move.' *Proceedings of the National Academy of Sciences of the USA* 100(10), pp5863–6. 13 May 2003.

Mill, J.S. *Utilitarianism* (1863). Charleston, BiblioBazaar, 2008.

Miller, D.W. *Popper Selections.* Princeton Paperback, 1985.

Millman, D. *Sacred Journey of the Peaceful Warrior.* HJ Kramer, 2004.

Morrell, Rima. *Travelling Magically*. Piatkus Books, 2008.

Murphy, Joseph. *The Power of Your Subconscious Mind.* Pocket Books, an imprint of Simon and Schuster, 2000.

National Gallery, London. http://www.nationalgallery.org.uk/paintings/joseph-wright-of-derby-an-experiment-on-a-bird-in-the-air-pump (cited 23 Feb 2013). Image and interpretation assisted in stimulating my postulate.

Nelson, C.A., de Haan, M., Thomas, K.M. *Neuroscience of Cognitive Development.* John Wiley & Sons, 2006.

Nietzsche, F. *Beyond Good and Evil.* New York, Oxford University Press, 2008.

Oerter, Robert. *The Theory of Almost Everything: The Standard Model, the unsung triumph of modern physics.* Plume, 2006.

Ophardt, C.E. 'Citric acid cycle summary.' 2003. Available at: http://www.elmhurst.edu/~chm/vchembook/612citricsum.html (accessed July 2013).

Oxfam. *Food Security: An Oxfam Perspective. Theory and practice of analysis in emergencies*. Oxford, Oxfam, 1997 (draft).

*Oxford Companion to Philosophy*. Ed. Ted Honderich. Oxford University Press, 1995.

Paarlberg, P. 'The weak link between world food markets and world food security.' *Food Policy* 25, pp317–55. 2000.

Pearce, D. *The Hedonistic Imperative.* http://www.hedweb.com/confile.htm (cited 2009).

Popper, Karl. Objective Knowledge: *An evolutionary approach.* Oxford,

Clarendon Press, 1972.

Popper, Karl. *In Search of a Better World: Lessons and essays from thirty years*. Translated with contributions by M. Mew. London, Routledge, 1st Edition, 1992.

Popper, Karl. *Knowledge and the Mind-Body Problem: In defence of interactionism*. Ed. M.A. Notturno. London, Routledge, 1994.

Popper, Karl. *The Logic of Scientific Discovery.* London, Routledge Classics, 2nd Edition, 2002.

Popper, K.R., Eccles, J.C. *The Self and Its Brain: An argument for interactionism*. Springer, 1st Edition, 1977.

Pulido, F. 'The genetics and evolution of avian migration.' *Bioscience* 57(2), pp165–74. 1 February 2007.

Quine, W.V. *Ontological Relativity and Other Essays*. New York: Columbia University Press.

Rachels, James M. *The End of Life: Euthanasia and morality*. Oxford University Press, 1986.

Rachels, James M. *Created from Animals: The moral implications of Darwinism*. Oxford University Press, 1990.

Riedl, R. (1984). *Biology of knowledge: The evolutionary basis of reason*. Trans. P. Foulkes. New York, Wiley, 1984 (originally published 1980).

Rogers, W., Hutchison, K. 'Evidence-based medicine in theory and practice: epistemological and normative issues.' 2015. Cited online via Springer link.

Roof, Judith. *The Poetics of DNA*. University of Minnesota Press, 2007.

Rozendaal, R.M., Koes, B.W., et al. 'Effect of glucosamine sulfate on hip osteoarthritis: a randomized trial.' *Annals of Internal Medicine* 148(4), pp 268–77. 19 February 2008.

Sackett, D.L., Rosenberg, W.M.C., Muir Gray, J.A., Haynes, R.B., Richardson, W.S. 'Evidence based medicine: what it is and what it isn't.' BMJ 312(7023), p71. 13 Jan 1996. doi: https://doi.org/10.1136/bmj.312.7023.71

Schelling, W.J. *Ages of the World*. http://plato.stanford.edu/entries/

schelling/#4 (cited Nov 2014).

Schlick, M. 'Die Kausalität in der gegenwärtigen Physik.' *Die Naturwissenschaften 19*, p156. 1931.

Schmoll, M., Yian, C., Sun, J., Tisch, D., Glass, N.L. 'Unravelling the molecular basis for light modulated cellulase gene expression: the role of photoreceptors in Neurospora crassa.' BMC Genomics 13, p127. 31 Mar 2012. doi: 10.1186/1471-2164-13-127. (Free PMC article, cited 14 Feb 2019)

Schumm, Bruce A. *Deep Down Things: The breathtaking beauty of particle physics*. Johns Hopkins University Press, 2004.

Seaman, J. 'Management of nutrition relief for famine affected and displaced populations.' *Tropical Doctor* 21, suppl. 1, pp38–42. 1991.

Singh, G. 'Recent considerations in nonsteroidal anti-inflammatory drug gastropathy.' *American Journal of Medicine*, p31. July 1998.

Spackman, K. The Ant and the Ferrari. HarperCollins, 2012.

Sprenger, M. 'Issues at the interface of general practice and public health: primary health care and our communities.' General Practice Online. 2005. Available at: http://www.priory.com/fam/gppublic.html (accessed Nov 2012).

Steen, R.J. 'Retractions in the scientific literature: is the incidence of research fraud increasing?' *Journal of Medical Ethics* 37(4), pp249–53. April 2011.

Tylor, Edward B. *Primitive Culture: Researches into the development of methodology, philosophy, religion, language, art and custom.* 1871. http://branchcollective.org/?ps_articles=peter-logan-on-culture-edward-b-tylors-primitive-culture-1871

Varela, F.J., Thompson, E., Rosch, E. *The Embodied Mind: Cognitive science and human experience*. MIT Press, 1st Edition, 1993, pp195–9.

Walsch, N.D. *Conversations with God*. Griffin Press, 1999.

Wilkens, P., Scheel, I.B., Grundnes, O., Hellum, C., Storheim, K. 'Effect of glucosamine on pain-related disability in patients with chronic low back pain and degenerative lumbar osteoarthritis: a randomized controlled trial.' *JAMA* 304(1), pp93–4. 7 July 2010.

Woodhandler, S., Himmelstein, D. *British Medical Journal* 345, pp50–1. 2012.

World Debt Clock. http://www.nationaldebtclocks.org/debtclock (cited Oct 2016).

World Health Organization. *Management of Severe Malnutrition: A manual for physicians and other senior health workers.* Washington, D.C. (abstract), 1999.

Wright, Robert. *The Moral Animal.* Little, Brown and Company, 1995.

Wuketits, Franz M., *Biologie und Kausalität: Biologische Ansätze zur Kausalität, Determination und Freiheit.* Verlag Paul Parey, 1981.

Wuketits, Franz M. '*Zum Konzept der inneren Selektion: Stellungnahme zu einer evolutionstheoretischen Kontroverse.*' Paläontologische Zeitschrift 59(1), pp35–41. June 1985 (cited 13 Sept 2014). doi: 10.1007/BF02985997

Wuketits, Franz M. *Evolutionary Epistemology and Its Implications for Humankind.* State University of New York Press, 1990.

Wuketits, Franz M., Antweiler, Christopher. *Handbook of Evolution. Vol. 1: The Evolution of Human Societies and Cultures.* John Wiley & Sons, 2008.

# TRANSFORMATION

Transform your life, transform your world - Changemakers Books publishes for individuals committed to transforming their lives and transforming the world. Our readers seek to become positive, powerful agents of change. Changemakers Books inform, inspire, and provide practical wisdom and skills to empower us to write the next chapter of humanity's future.
If you have enjoyed this book, why not tell other readers by posting a review on your preferred book site.

## Recent bestsellers from Changemakers Books are:

**Integration**

The Power of Being Co-Active in Work and Life

Ann Betz, Karen Kimsey-House

*Integration* examines how we came to be polarized in our dealing with self and other, and what we can do to move from an either/or state to a more effective and fulfilling way of being.

Paperback: 978-1-78279-865-1 ebook: 978-1-78279-866-8

**Bleating Hearts**

The Hidden World of Animal Suffering

Mark Hawthorne

An investigation of how animals are exploited for entertainment, apparel, research, military weapons, sport, art, religion, food, and more.

Paperback: 978-1-78099-851-0 ebook: 978-1-78099-850-3

**Lead Yourself First!**

Indispensable Lessons in Business and in Life

Michelle Ray

Are you ready to become the leader of your own life? Apply simple, powerful strategies to take charge of yourself, your career, your destiny.

Paperback: 978-1-78279-703-6 ebook: 978-1-78279-702-9

**Burnout to Brilliance**

Strategies for Sustainable Success

Jayne Morris

Routinely running on reserves? This book helps you transform your life from burnout to brilliance with strategies for sustainable success.

Paperback: 978-1-78279-439-4 ebook: 978-1-78279-438-7

**Goddess Calling**

Inspirational Messages & Meditations of Sacred Feminine Liberation Thealogy

Rev. Dr. Karen Tate

A book of messages and meditations using Goddess archetypes and mythologies, aimed at educating and inspiring those with the desire to incorporate a feminine face of God into their spirituality.

Paperback: 978-1-78279-442-4 ebook: 978-1-78279-441-7

**The Master Communicator's Handbook**

Teresa Erickson, Tim Ward

Discover how to have the most communicative impact in this guide by professional communicators with over 30 years of experience advising leaders of global organizations.

Paperback: 978-1-78535-153-2 ebook: 978-1-78535-154-9

**Meditation in the Wild**

Buddhism's Origin in the Heart of Nature

Charles S. Fisher Ph.D.

A history of Raw Nature as the Buddha's first teacher, inspiring some followers to retreat there in search of truth.

Paperback: 978-1-78099-692-9 ebook: 978-1-78099-691-2

**Ripening Time**

Inside Stories for Aging with Grace

Sherry Ruth Anderson

*Ripening Time* gives us an indispensable guidebook for growing into the deep places of wisdom as we age.

Paperback: 978-1-78099-963-0 ebook: 978-1-78099-962-3

**Striking at the Roots**

A Practical Guide to Animal Activism

Mark Hawthorne

A manual for successful animal activism from an author with first-hand experience speaking out on behalf of animals.

Paperback: 978-1-84694-091-0 ebook: 978-1-84694-653-0

## What people are saying about

# Spheres of Perception

Moving beyond and between disciplines and the effects of technology on our lives, this book provides a rich and sophisticated transdisciplinary exploration of humanity's 'being in this world.' The reflections on our logical, physical, and metaphysical evolution challenge our illusions about humanity's competence to overcome disparities between the way we live and the way we develop. This book must be read by everybody looking for a sensible and holistic evaluation of the drastic challenges we face and the transformations we require to adapt to the present.
**Dr Hester du Plessis,** DLitt et Phil, Chief Research Specialist, Human Sciences Research Council (HSRC), South Africa

# Spheres of Perception

## Morality in a Post Technocratic Society

Theodore Holtzhausen